THE BOOK OF
THE CROSSBOW

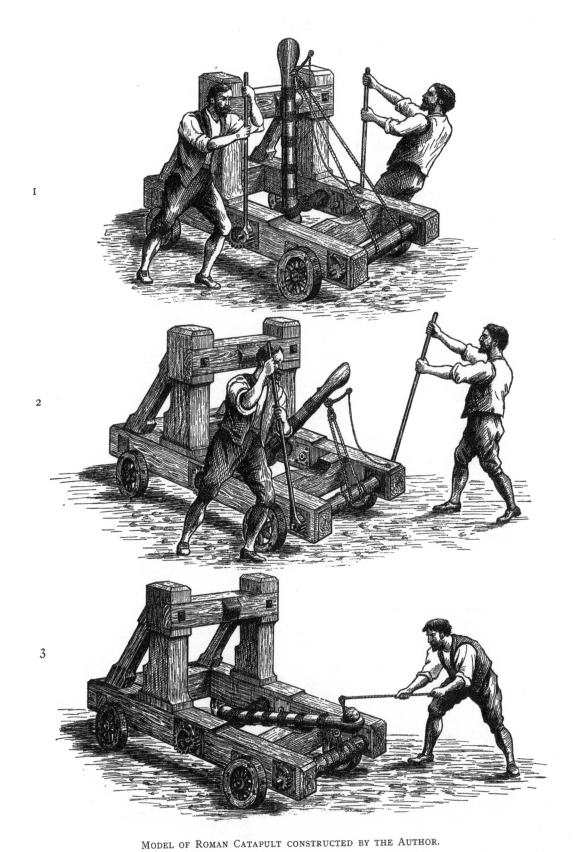

MODEL OF ROMAN CATAPULT CONSTRUCTED BY THE AUTHOR.

Weight, 1½ tons.   Range, with a 6-lb. stone ball, 300 yards.

1. Twisting up the skein of cord by means of the winches.   2. Winding down the arm.
3. Releasing the arm when fully wound down.

THE BOOK OF
THE CROSSBOW

Ralph Payne-Gallwey

DOVER PUBLICATIONS, INC.
New York

Bibliographical Note

This Dover edition, first published in 1995, is an unaltered and unabridged repub-
lication of the work originally published by Longmans, Green, and Co., London, in
1903 under the title *The Crossbow / Mediæval and Modern / Military and Sporting / Its
Construction History and Management / With a Treatise on the Balista and Catapult of
the Ancients*. The Dover edition also includes the author's Appendix, published separately [no publisher given], in 1907 under the title *Appendix to the Book of the Crossbow
and Ancient Projectile Engines*.

Library of Congress Cataloging-in-Publication Data

Payne-Gallwey, Ralph, Sir, 1846–1916.
    The book of the crossbow / Ralph Payne-Gallwey.
        p.   cm.
    Originally published: The crossbow. London : Longmans, Green, and Co., 1903.
    Includes index.
    ISBN-13: 978-0-486-28720-1 (pbk.)
    ISBN-10: 0-486-28720-3 (pbk.)
    1. Crossbows. 2. Catapult. I. Payne-Gallwey, Ralph, Sir, 1846–1916. Crossbow. II.
Title.
U878.P38 1995
623.4′41—dc20
                                        95-22140
                                          CIP

Manufactured in the United States by LSC Communications
28720308   2018
www.doverpublications.com

PREFACE

In this volume I have endeavoured to describe the history mechanism and manipulation of the crossbow; mediæval and modern, military and sporting.

Though there are numerous books essays and manuscripts which deal exhaustively with the longbow, the subject of the crossbow has not previously been treated, other than in a very cursory manner by writers on the armour and weapons of the Middle Ages.

I cannot, in any language, discover a work exclusively devoted to the crossbow, though this arm was carried by hundreds of thousands of soldiers in mediæval warfare, and has ever since been popular on the Continent for sporting or target use.

In the fourteenth and fifteenth centuries the longbow was the cherished weapon of the English, while the crossbow held a similar position in France Germany Italy and Spain.

The longbow, glorious as its achievements were in the hands of our ancestors, was but a hewn stick of foreign yew of no intrinsic value.

On the other hand, the crossbow gave the artist, the engraver, the inlayer and the mechanic every chance of exercising their talents to the utmost.

There are but one or two old English longbows in existence; there are, however, numbers of beautifully constructed mediæval crossbows to be seen in armouries and museums; weapons which were originally made with as much skill and nicety as a costly modern gun.

I have added a treatise on the great projectile engines of the ancients, as they have not before been fully described or criticised.

As some of the ancient siege engines resemble a crossbow and are supposed to have suggested its invention, I trust the details I give of the history and construction of these curious machines will be of sufficient interest to justify their inclusion.

THIRKLEBY PARK,
THIRSK.

CONTENTS

CONTENTS

PART III

THE CONSTRUCTION AND MANAGEMENT OF CROSSBOWS—(*continued*) : MODERN

Part IV

A TREATISE ON THE SIEGE ENGINES USED IN ANCIENT AND MEDIÆVAL TIMES FOR DISCHARGING GREAT STONES AND ARROWS

APPENDIX (separately paginated)

ILLUSTRATIONS

Part I

THE HISTORY OF THE CROSSBOW

PART II

THE CONSTRUCTION AND MANAGEMENT OF CROSSBOWS: MEDIÆVAL

PART III

THE CONSTRUCTION AND MANAGEMENT OF CROSSBOWS (continued): MODERN

Part IV

ANCIENT AND MEDIÆVAL SIEGE ENGINES

ILLUSTRATIONS

NOTE BY AUTHOR

The Crossbows and their details I have drawn chiefly from examples in my own collection, and the working plans of siege engines from large and small models which I have constructed.

The Pictorial Illustrations from Mediæval books and Manuscripts have been most admirably copied from the originals—for subsequent reproduction herein—by Mr. W. Woodrow of the British Museum Library.

HUNTERS WITH CROSSBOWS.

From Tempesta, Antonio, a Florentine painter, b. 1555; d. 1630.

Part I

THE HISTORY OF THE CROSSBOW

WITH NOTES COMPARATIVE ON

THE LONGBOW, SHORTBOW, AND HANDGUN

THE NOMENCLATURE OF THE CROSSBOW.

The MEDIÆVAL CROSSBOW was called by many names, most of which were derived from the word Balista. The Balista was a great siege engine on wheels that was used by the ancients, and which in appearance and mechanism resembled a crossbow, though so much larger in size, Chapter LVII.

THE CROSSBOWMAN WAS KNOWN AS [1]—

Arbalista	Arcubalistarius
Arbalistarius	Arcubalistus
Arbalistator	Balistarius
Arbalistanus	Balistrarius
Arbalistrius	Balistrator

THE CROSSBOW—

Arbalet	Arblast
Arbalist	Alablaste
Arbalista	Alblast
Arbaliste	Arbelaste
Arcubalist	Arowblaste
Arcubalista	Arblat
Arcubalistus	Arbalestel [2]
Manu-balista	

CROSSBOW SHOOTING—

Arbalestry	Alblastrye

THE NAME OF THE CROSSBOW IN DIFFERENT COUNTRIES AT THE PRESENT DAY:

France ⎫
Belgium ⎬ Arbalète

Italy—Baléstra

Spain—Ballésta

Portugal—Bésta

Germany ⎫
Austria ⎬ Armbrust

Denmark—Flitsbue

Sweden—Armbost

Norway—Krydsbue

Russia—Samostrel

---

[1] To avoid needless repetition, the names are here spelt with an 'i,' as Arbalist. They were, however, just as commonly spelt with an 'e,' as Arbalest. It will be seen that in some cases the same word stands for a Crossbowman and a Crossbow.

In Mediæval English the Crossbowman was known as Alblaster, Alblastere, Allblawster, Arbalaster, Arbalister, Arblaster, Arowblaster, Awlblaster; all these words being corruptions of the Latin.

[2] In old French, a small crossbow.

CHAPTER I

THE MILITARY CROSSBOW

FIG. I.—NORMAN CROSSBOWMEN.

From Manuscript of Matthew Paris.[1]

THE CROSSBOW was, probably, introduced into England as a military and sporting arm by the Norman invaders in 1066.

Early in the twelfth century, the construction of this weapon, the bow of which was not yet formed of steel, was so much improved that it became very popular in both English and Continental armies.

The wounds caused by the crossbow in warfare were, however, considered so barbarous, that its use, except against infidels, was interdicted by the second Lateran Council, in 1139, under penalty of an anathema, as a weapon hateful to God and unfit for Christians. This prohibition was confirmed, at the close of the same century,. by Pope Innocent III. Conrad III. of Germany, 1138–1152, also forbad the crossbow in his army and kingdom.

The employment of crossbowmen, nevertheless, again became common in English and Continental armies in the reign of Richard I., 1189–1199, and the death of this king, which was caused by a bolt from a crossbow, (at the siege of the Castle of Chaluz, near Limoges, in France, in 1199,) was thought to be a judgment from Heaven inflicted upon him for his disobedience and impiety in permitting crossbowmen to enter his service.

Richard was an expert with the weapon. At the siege of Ascalon—though prostrated with fever—he is said to have been carried from his tent on a mattress, so that he might enjoy the pleasure of shooting bolts at the defenders of the town. In this case, however, as the enemy consisted of Turks and infidels, his act would have been sanctioned by the Church of Rome. Though among

---

[1] As the period in which a chronicler, or artist, lived, cannot be repeated on every occasion when his name occurs, consult, therefore, the general index to ascertain any biographical date.

English soldiers, the longbow began to supersede the crossbow and the shortbow, during the reign of Edward I. in the last few years of the thirteenth century, crossbows continued to be held in some favour in our armies. In the list of troops mustered by Edward II., in 1319, for the siege of Berwick, crossbowmen are enumerated as part of the forces. In Scotland and Ireland, the crossbow was almost unknown, and even the bow was sparingly used, though in Wales, as in England, the latter was the common arm of the people in the fourteenth and fifteenth centuries.

For about two centuries and a half (1200–1460) the crossbow was the favourite weapon on the Continent. It was almost equally popular with English commanders and soldiers till about 1290, and several estates in this country were held by the service of delivering a crossbow when the king passed through them.

FIG. 2.—CROSSBOWMEN.

The soldiers carry windlass crossbows. One man is winding up his weapon; the other is shooting, with his windlass laid on the ground at his feet.

From Manuscript, Froissart's ' Chronicles.'

The Genoese were always famed for their skill in the construction and management of crossbows, and were hired for service by sea and land by all nations on the Continent. They are said to have used these weapons with success, even as early as 1099 at the siege of Jerusalem. In the naval engagement near Sluys, in Holland, where Edward III. defeated the French in 1340, the latter had as many as 20,000 Genoese crossbowmen on their ships, and the largest numbers of crossbowmen ever seen in order of battle on land, were probably the 15,000 Genoese who, according to Froissart,[1] formed the front rank of the French army at Crécy in 1346. It is asserted by numerous historians, all of whom derive their information on the subject from a cursory statement by the second continuator of William of Nangis,[2] that the crossbowmen at Crécy were unable to shoot with effect, because the strings of their weapons were slack owing to the great storm of rain that set in just before the battle. Muratori,[3] the Italian antiquary, declares that the

[1] Sir John Froissart—French chronicler, born about 1337, died about 1410.
[2] William of Nangis—French historian, a Benedictine monk of the Abbey of St. Denis, flourished in the thirteenth century, wrote a history of the kings of France, died 1300.
[3] Muratori—Italian priest and historian, born 1672, died 1750.

ground at Crécy was so boggy that the crossbowmen could not stand firm when they endeavoured to stretch the strings of their weapons; but as the field of Crécy consists of rather steep downs, and not of lowland, it is not probable that the state of the ground impeded the crossbowmen.

Although much doubt has been thrown on the statement that the crossbows of the Genoese failed to act on this occasion, owing to their strings being slackened by wet weather, it is possible that the incident occurred, without, however, in any measure influencing the result of the battle.

The strings might easily have been rendered less effective than usual by the heavy rain that fell just before the battle, and by the bright sun which is known to have succeeded the rain.

This combination of water and heat would certainly relax in some degree the strings of the crossbows used at the time of Crécy, if they were uncovered, and would make the strings too loose to be of good service, till they could be removed from the bows in order to be shortened by twisting, and then replaced; all of which would entail, of course, time and care.

It should be remembered that the bows of the Genoese crossbowmen at Crécy were doubtless composite ones, made of wood, horn, sinew, and glue, bows of steel being of later introduction.

The composite bow was straight, hence its bow-string was fixed to it in a necessarily rather slack condition; for this reason the threads composing its string, being more or less detached, were liable to absorb moisture.

On the other hand, the threads that composed the tightly strained string of a steel crossbow, lay closely packed together, and as in this case the string was always thickly smeared, both inside and outside, with beeswax to preserve it, it was impervious to water.

To test the matter, I have sunk a steel crossbow in a tank of water for a day and a night and have found no appreciable alteration in the tightness of its string. I have also placed in water a crossbow with a comparatively loose string—such as those which I believe were used by the Genoese at Crécy—and found that after half an hour's submersion, the application of a lever to bend the bow caused the string subsequently to stretch down the stock an inch further than its proper position, its tautness, and consequent effectiveness, thus being lost.

The supposition that the crossbows of the Genoese at Crécy had bows of wood, or of wood and horn, is confirmed in a curious way by David-ap-Gwilym, a famous Welsh bard and archer of the fourteenth century. In one of his poems, the bard refers to a soldier who had sailed with Edward III. to fight at Crécy, and whom he had cause to hate, as he had supplanted the poet in the affections of his mistress. The poet calls upon the enemy to shoot his more fortunate rival, with the 'arbalest' or short stirrup stick. The translation of this passage, as rendered by A. J. Johns (the italics are mine), runs:

> And thou crossbowman true and good,
> *Thou shooter with the faultless wood,*
> Haste with thy stirrup-fashioned bow
> To lay the hideous varlet low.

As further proof that at Crécy the Genoese did not use the powerful steel crossbow which was bent by a windlass, I quote the following extract from Viollet-le-Duc (Dictionnaire raisonné du Mobilier français. Paris 1868–75). 'John II., King of France (the Good), issued in 1351 a military regulation which ordered that the crossbowman who had a good crossbow, strong according to his strength, should receive three sous tournoise wages per day.' This plainly shows that the military crossbow of the time of Crécy was bent either by the hands alone, or, as was more probable, by a thong and pulley, a claw fixed to the girdle, or by means of a goat's-foot lever. If the crossbowmen referred to in the regulation given above had steel crossbows with windlasses, such as were commonly used towards the end of the century, the question of regulating the power of the bow to the strength of the soldier would not have arisen, as with a windlass a boy could bend the thickest of steel bows.

The Genoese at Crécy (they were in the first line and were the only troops of the French army who advanced towards the English in fair order) were probably checked, and thrown into confusion, by showers of arrows, before they could approach their assailants sufficiently near to discharge one crossbow bolt with effect.[1] All contemporary and later evidence tends to prove, that the crossbows carried by the Genoese at Crécy had not steel bows; thus they could not compete at all with the English longbow, as they had formerly done with the old shortbow.

The Genoese became, therefore, a large and helpless target for the English bowmen, and very soon scattered and fled, for they were unable to inflict any loss upon their opponents, though struck down in numbers themselves.

This, in itself, was sufficient to throw these unfortunate mercenaries into a state of panic, even had their small crossbows been in proper condition, as indeed they may have been, notwithstanding tradition and surmise to the contrary.

When the crowding mass of horse and foot, which for several miles had been pressing in disorder on the heels of the Genoese, came up, they found the crossbowmen in hot retreat, either by reason of the deadly hail of English arrows they had just encountered, or because of the uselessness of their weapons.

The cavalry, however, in merciless manner, galloped furiously over the

---

[1] It is probable that the crossbows carried by the Genoese at Crécy were unable to send their bolts further than about 200 yards.

luckless crossbowmen and hewed them down with their swords, as cowardly knaves whose broken ranks blocked the way to the front. Whether the alleged incident of the crossbow-strings occurred or not, or whether it was said by the Genoese to have taken place as an excuse for their discomfiture, we shall never know. At all events one thing is certain, and that is, that at the time of Crécy the longbow must have excelled considerably the crossbow in range and penetration.

Even when the powerful steel crossbow with its windlass was invented, it was rightly considered to be less efficient in open warfare than the longbow, which was light, portable, and inexpensive, and could be discharged five or six times to the crossbow's once.

Whilst the crossbowman was occupied in stretching the string of his bow, the archer with a longbow could be assailing him with a succession of arrows.

For this reason, the crossbowman was often attended in battle by a companion, who sheltered him from the arrows of the enemy by holding before him a thick shield of wood and hide, whilst he was pulling up his bow-string.[1]

Sometimes the crossbowman carried a small shield himself, which he slung on his back on the march, and propped up before him as a protection when shooting, or when bending his crossbow.

FIG. 3.—CROSSBOWMEN.

The centre figure is winding up his windlass crossbow behind the shelter of a shield.

From Manuscript, Froissart's 'Chronicles.'

The crossbow may be described as the blunderbuss of archery, and the larger sort was much employed in the defence of fortresses, as behind the shelter of turrets and loopholes a heavy crossbow could be conveniently rested, and the weapon could then be aimed in safety at a besieging force. It was also a favourite weapon on board ships of war.

It was certainly superior to the longbow in some respects; for besides its much heavier missile, and its accuracy and power as an instrument of offence or defence in fortified positions, it could be used from any position of concealment demanded by the exigencies of war, as, for instance, through the peepholes and slits of low basement rooms, or through the small loopholes that were pierced in the walls of the flanking towers of a fortification to enfilade

---

[1] The larger shields, which were carried before the knights (by their pages) when on the march, and which were propped up in front of them as a protection from arrows in a battle or a siege, were known as pavises or mantlets.

FIG. 4.—A SHIP OF WAR, WITH CROSSBOWMEN.

From Valturius.—Edition 1472.

[Of this plate Valturius quaintly writes: 'When everything is cleared for navigation before the charge is made upon the enemy, it is well that those who are about to engage the foe should first practise in port, and grow accustomed to turn the tiller in calm water, to get ready the iron grapples and hooked poles, and sharpen the axes and scythes at their ends. The soldiers should learn to stand firm upon the decks and keep their footing, so that what they learn in sham fight they may not shrink from in real action.']

the approach to its gateway. A crossbow could be strung in, and discharged from, a room not 6 feet high to the ceiling, whilst a longbowman required a height of at least 7 feet in order to shoot an arrow with effect.

Nor did the crossbow require the strength, skill, and practice to manipulate it that were so necessary in the case of the longbow.

The narrow cruciform loophole, called by architects 'Arbalestina,' which is usually to be seen in the masonry of a mediæval fortress, was designed for the special use of crossbowmen in repelling an assault.

To enable the crossbow, or longbow, to be aimed to the right or left through a loophole, the aperture was greatly widened out on the inside face of the perforated wall.

FIG. 5.
ARBALESTINA
From a Glossary of Terms of Architecture, 1840.

The perpendicular loopholes, also common in ancient castles, were intended for the archer with his longbow, hence they were not cruciform in outline.[1]

The perfected military crossbow of the fifteenth century, with its steel bow and appendages, being heavy, and slow in action, could not be utilised so readily for shooting quickly at single combatants, or at small bodies of men and horse on the open field of battle, as could the longbow. Its weight alone precluded it from being aimed with success against rapidly moving objects, nor could its bolt be directed with precision if a hurried aim was taken.

On the other hand, a skilful archer with his longbow might quite possibly pierce a galloping stag with an arrow at a distance of 70 yards, and, if he failed to strike his mark, send another shaft at his quarry before it was out of bow-shot.[2]

This advantage of rapid aiming and shooting, the longbowman could apply

FIG. 6.—CROSSBOWMEN

They represent French soldiers at the defence of Rouen, 1419, shooting from behind the shelter of shields propped up in front of them.

From Cotton Manuscript.

[1] 'Our Château de Cheeignee we have assigned to the Earl of Montfort in such wise that he is to understand we cannot allow in it any perpendicular loophole for archers, nor any cruciform loophole for crossbowmen.'—From a Royal Charter of France dated 1239 and quoted in Sir S. Meyrick's work on Ancient Armour.

[2] If an archer expected to use two arrows in rapid succession, he held his second arrow against the back of his bow with his left hand, or else pressed into the palm of the right hand by the thumb, so that he could instantly seize it and fit it to his bow-string, and thus save the time that would otherwise be spent in extracting it from a quiver. On the other hand the crossbowman, when bending his bow, held a bolt between his teeth, so that it might be ready to fit to his weapon without any delay. Pages 49, 124.

equally to the destruction of a single enemy, at the moderate range—to the archer—of 80 to 100 yards, or at much longer distances in the case of groups of horse or foot.

It will be understood from what I have written that its weight and size, and tedious manipulation, were the drawbacks of the crossbow in open battle, and that its heavy bolt, great power and accuracy, and its convenience for the defence or attack of fortifications, were its advantages.

CHAPTER II

THE SPORTING CROSSBOW

FIG. 7.

HOW A CROSSBOWMAN SHOULD APPROACH ANIMALS BY MEANS OF
A CART CONCEALED WITH FOLIAGE.

From Manuscript of Gaston Phœbus. Fourteenth century.

THOUGH the SPORTING CROSSBOW, as was the case with the crossbow employed in warfare, never found as much favour in England as it did on the Continent, it was in limited use among the nobles and gentry of the kingdom for killing deer, the extreme accuracy of the weapon at a short range, and the heavy bolt it threw, well adapting it for the chase.

The crossbow, moreover, could be used by the hunter as he crouched behind trees or rocks, or amid the dense cover that formerly compassed the haunts of deer, in places where the string of a longbow could not be fully drawn for want of space, and when the act of doing so, were it possible, would probably alarm and drive away the animal for which the hunter was lying in ambush. The hunter could carry his crossbow ready bent, and then discharge it from any position, even when lying on the ground, while the archer with a longbow could not shoot with effect from a stooping or recumbent attitude.[1]

The crossbow was also noiseless as well as powerful and accurate, and for this reason it survived—as a common weapon of chase—the first serious introduction of the hand-gun for over a century and a half—1470–1630.

[1] I find that the thick steel bow of the ancient military, or sporting crossbow, like the spring of a gun-lock, does not 'tire'—*i.e.*, lose any of its power—even though it be kept bent for two or three hours at a time

A heavy steel bow was slightly bent in proportion to its length, and differed in this respect from the much lighter bow of a modern sporting crossbow. The latter is always liable to take a slight 'set,' or permanent bend, if kept in a strained condition for more than about ten minutes.

The time and money lavished on the ornamentation of high-class sporting crossbows, especially those of late sixteenth-century Continental manufacture, were very considerable, the best workers in metal, ivory and mother of pearl, being employed in their decoration.

The stock of the sporting crossbow was often covered with artistic representations of animals, birds and hunting scenes, surrounded by scroll-work, all finely chased and inlaid in silver, ivory and pearl.

The polished metal fittings of the stock, and even the hardened surfaces of the steel bow, were sometimes deeply inlaid with a delicate tracery in gold of leaves and flowers, or heraldic designs.

FIG. 8.

CROSSBOWMAN APPROACHING GAME BY MEANS OF A STALKING HORSE.

From MS. of Gaston Phœbus. Fourteenth century.

Different workmen constructed the distinct parts of a good sporting crossbow, just as the separate pieces of a gun are treated in these days by various artisans, before they are fitted together to produce the weapon in its finished state.

One set of craftsmen made the stock, another the windlass or the cranequin, and so it was with the lock and the string ; but the most important artificers of all were the men who forged and shaped the steel bows. The bows from Mondragon in Spain, which were of the same quality of steel as that of the famous Toledo sword blades, and those from Pyrmont in Germany, were celebrated for their excellence of strength and temper.

In confirmation of this we read in Sir J. Harington's translation of Ariosto (Italian poet, 1474–1533) :

> But as a strong and justly tempered bow
> Of Pyrmont steel, the more you do it bend,
> Upon recoil doth give the bigger blow,
> And doth with greater force the quarrel send.
>   (*Orlando Furioso.*)

The sporting crossbow of the sixteenth century, or from about 1500 to 1630, was no doubt a very effective weapon in its day for the purposes for which it was required, as the experience and skill of several centuries had brought it to perfection, ere it was at length superseded by the improved arquebus.

The hunter could not, however, bring down birds on the wing with his crossbow ; nor, indeed, could the man who used the arquebus of the same period, its system of ignition being so slow and primitive. The utmost the crossbowman could do was to lodge a bolt, often, in foreign countries, a poisoned one,[1] in the head or heart of a deer, bear, or wolf, standing, or passing slowly within about sixty paces; or else, perhaps, tumble over a crane or heron perched on the top of a tree.

In the time of crossbows, and early hand-guns, it should be remembered that deer and other animals were tame and easily stalked, and that wildfowl and game-birds were chiefly taken in nets and snares, and with trained hawks.

[1] From the practice of formerly steeping the heads of crossbow bolts in the juice of a poisonous herb, the white hellebore is to this day known in parts of the country districts of Spain as 'the crossbowman's plant.'

CHAPTER III

THE GENERAL DIMENSIONS OF CROSSBOWS

THE formidable SIEGE CROSSBOW of about 18 lbs. weight, which was only employed in the attack or defence of a fortress, though it could be supported and aimed by a man of very strong physique, was usually discharged either as it rested on a parapet, or when pivoted on a small tripod.

Not long since I was fortunate in obtaining from Nuremberg a fine example of one of these large weapons.

The woodwork of its stock was naturally much damaged by age and neglect, and this, and the lock and other fittings, I found it necessary to renew ; all of which I had carefully done by mechanics in my workshop. The steel bow is, however, the original one and of as good temper as ever, though it was made in Genoa over four hundred years ago.

The bow is 3 ft. 2 in. long, and at its centre $2\frac{1}{2}$ in. wide and 1 in. thick.

Shooting this crossbow from the shoulder, with a bolt 3 oz. in weight and 14 in. in length, I have attained a range of 460 yards, and at 60 yards I have sent a bolt right through a deal plank $\frac{3}{4}$ in. thick.

By suspending the crossbow in a perpendicular position from a beam, and then attaching heavy weights to a rope fastened to the centre of its bowstring, I was able to determine its strength of pull. The total weight required to draw the string of its bow 7 in., or from a state of rest to the catch of the lock, is 1,200 lbs. or over half a ton ! This, of course, gives the power of the crossbow in question, just as 50 lbs. represents the strength of an ordinary longbow, or the weight required to draw its string the length of its arrow.

Notwithstanding its immense strength of pull, by the aid of its portable little fifteenth-century windlass, the string of this crossbow can be stretched to the catch of its lock by the fingers of one hand, showing the great power and cleverly designed efficiency of the windlass of a mediæval crossbow.

It is, perhaps, worth recording here, that in the autumn of 1901 I shot several bolts with this weapon across the Menai Straits, from the battery of Fort Belan to Abermenai Point ; this was done in the presence of a number of sporting friends who were interested in the attempt, and who declared that the feat was impossible.

The distance achieved by the bolts, according to Ordnance Survey, was between 440 and 450 yards.

It is most unlikely that a missile of any kind has previously been projected without the aid of gunpowder, from one shore to the other, across this arm of the sea.

The large MILITARY CROSSBOW with a thick steel bow, which was carried by the crossbowman in battle, as at Agincourt for instance, weighed from 15 lbs. to 16 lbs. without its windlass.

Its steel bow was from 2 ft. 7 in. to 2 ft. 8 in. long, and at its centre $1\frac{3}{4}$ in. to 2 in. wide, and $\frac{5}{8}$ in. to $\frac{3}{4}$ in. thick.

The SPORTING CROSSBOW for killing deer by means of an ordinary bolt, weighed from 12 lbs. to 14 lbs. without its windlass; or, by reason of its then smaller stock, from 10 lbs. to 12 lbs. if a cranequin, instead of a windlass, was employed to wind up the bow-string. Its steel bow was from 2 ft. 5 in. to 2 ft. 6 in. long, and at its centre $1\frac{1}{2}$ in. to $1\frac{3}{4}$ in. wide, and $\frac{1}{2}$ in. to $\frac{5}{8}$ in. thick.[1]

The SMALLER SPORTING CROSSBOW, such as was used in Spain for killing deer with a poisoned bolt, and small animals and large birds with an ordinary bolt, weighed from 8 lbs. to 9 lbs. without its cranequin. Its steel bow was from 2 ft. 4 in. to 2 ft. 5 in. long, and at its centre $1\frac{3}{8}$ in. to $1\frac{1}{2}$ in. wide, and $\frac{3}{8}$ in. to $\frac{1}{2}$ in. thick.

NOTE.—The details concerning dimensions, weights and ranges given in Chapters III. and IV. are derived from a careful personal inspection and trial of a large number of late fifteenth, and early sixteenth-century crossbows. The ranges were in all cases measured by surveyor's chain and not by foot-pace.

[1] This weapon is described in Chapters XIX.–XXVIII.

CHAPTER IV

THE BOLTS USED WITH CROSSBOWS

FIG. 9.—A STORE OF CROSSBOW BOLTS, SHAFTS, AND HEADS.
The crossbowman is aiming at a target to the left of the picture.

From a catalogue of the Arsenal of the Emperor Maximilian I. (b. 1459, d. 1519).

As the result of numerous experiments with crossbow bolts of different weights and shapes, I have found that bolts of seasoned yew, weighing from $2\frac{1}{2}$ oz. to $2\frac{3}{4}$ oz., and measuring about 12 in. in length by $\frac{1}{2}$ in. to $\frac{5}{8}$ in. in diameter, fly the straightest and furthest; bolts of these dimensions being similar to those to be seen in museums and intended for use in the military crossbow which had a strong steel bow.

The light bolts which I have tried, weighing from $1\frac{1}{4}$ oz. to $1\frac{1}{2}$ oz., did not in any instance, however true their flight, carry nearly so far as the heavier ones, the former never seeming to feel the full energy of the bow-string of a powerful crossbow.

Even bolts suitable for a large sporting crossbow require to be from $2\frac{1}{4}$ to $2\frac{1}{2}$ oz. in weight.   Fig. 77, p. 126.

In order to give a sufficient substance for the thick string of the crossbow to act against, the height of the bolt at its butt-end was the same as the diameter of the bow-string, which was usually $\frac{1}{2}$ in. thick.

The bolt was then tapered forward to a slightly increased size, which caused the fore-end of its shaft, over which the sheath of the iron head fitted, to be $\frac{5}{8}$ in. thick.

This gradual preponderance of weight towards its point was a necessary feature in the short length of a bolt for a crossbow, in order to give it a proper balance.

The fact of the bolt being tapered as described, also caused the head and butt only of the bolt to rest on the grooved stock of the crossbow; hence friction was reduced, and a longer flight given to the missile than would otherwise have been the case.

As the recovery of the bolts and arrows discharged in warfare could not well be expected, especially by the vanquished side, and as immense numbers were required, they were but of plain construction, for a bestowal of high finish would have been a waste of money and labour.

On the other hand the bolts and arrows intended for killing deer, or for target practice, were made with great nicety, their smoothly hammered steel points being shaped with particular care, and their shafts had always three feathers, the bolts and arrows used in war usually having only two.

Crossbow bolts for military service were often winged with thin strips of wood, leather, skin or horn, instead of with goose or swan feathers.

In some instances a bolt had its feathers, or the material that took the place of feathers, fixed on spirally, so as to cause the shaft to rotate, with a view to its maintaining a true direction in transit through the air.

This spinning bolt was called a 'Vireton,' from the French word 'Virer,' to turn, but I do not find it is more accurate than the usual kind, or has so long a flight.

A simple and effective method of winging crossbow bolts in former times was as follows. A fine slit was sawn a few inches up the centre of the shaft from its butt-end, an oblong piece of thin dry leather, or parchment, was next drawn tight into the slit, and the butt of the bolt was wrapped round with waxed thread close behind the material inserted, the latter being then trimmed to a proper outline.

The heads of military crossbow bolts were of solid metal, prolonged to a hollow sheath to fit over the wooden shaft. Some bolts were sharply

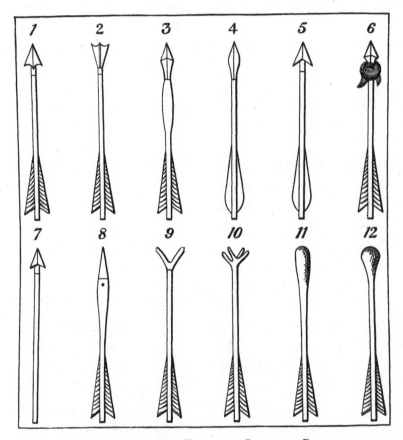

FIG. 10.—DIFFERENT FORMS OF CROSSBOW BOLTS.

1, 2, 3, 4, 5, 6, 7, military bolts ; 6, bolt with tow soaked in oil for firing ships and houses ; 7, bolt for a slur bow ; 8, bolt for killing deer ; 9, 10, bolts for killing large birds ; 11, 12, bolts for killing game birds. The latter had not metal heads, and were blunt, so as not to damage the game.

pointed, and these were intended for use against men in buff coats or in light armour, such as foot-soldiers.

Other bolts had square-faced heads with four small points, one at each corner of the head, so that they might not glance off armour, but give a straight and smashing blow to mounted men wearing breast-plates and helmets, against which the end of a sharp projectile might break, bend, or turn aside. 2, fig. 10.

From the shape of its head, usually four-sided, whether blunt or pointed, the bolt of the early military crossbow was called a quarrel,' a name afterwards applied to all crossbow bolts of whatever form.[1] The bolts of crossbows were not 'nocked' at their butt-ends like the arrows of a longbow. I have never seen any original crossbow bolts with nocks, and from experiments of my own I find that they fly further and with more accuracy when they have plain upright ends.

There is little doubt that a strong military crossbow, with a steel bow, was able, at a fair range, to penetrate with a sharp-headed bolt any armour that was worn at the time of the introduction of this weapon into warfare, though the arrow of a longbow could not always be depended on to do so, as its shaft was more apt to break on contact.

The bolt of a steel crossbow was a heavy and dangerous projectile, even when at the end of its flight ; far more so than an arrow. A bolt which did not happen to penetrate armour, must have dealt a blow sufficient to stun a man and knock him to the ground.

Even the primitive hand-gun, at close range, would probably shiver with its ball the mail and armour worn at the period when this weapon came into use.

Armour was most likely first increased in thickness, and plate generally substituted for chain mail, in order to resist the sharp head of the arrow of the powerful longbow. When the hand-gun was introduced, and the crossbow with a thick steel bow perfected, breast-plates, helmets, and other metal protections for horse and man, were still further strengthened : so much so, that near the end of the fifteenth century they attained a weight almost beyond the physical endurance of the knights on horseback or the men-at-arms.

Finally plate armour was much lightened, and to a considerable extent discontinued, in the reign of Henry VIII., as being of small avail against gunpowder and bullets, though it was worn in its less complete form till near the end of the seventeenth century as a protection from sword and pike.

Though the knights, secure in their heavy armour, had no scruples in riding down and killing the leather-clad foot-soldier, it is entertaining to read of the fierce outcry they made when the foot-soldier retaliated with steel crossbow and arquebus.

The knights called Heaven to witness that it was not honourable warfare to employ such weapons in battle, the fact being that they realised that armour was no longer the protection to their persons which it was before the days of heavy crossbows and hand-guns.

[1] Probably from *quarreau*,—in modern French *carreau*, anything diamond-shaped or with squared faces, as was usually the head of a crossbow bolt. 3, fig. 10.

CHAPTER V

THE RANGE OF THE MEDIÆVAL CROSSBOW AND HOW IT COMPARED IN THIS RESPECT WITH THE LONGBOW

FIG. 11.—CROSSBOWMEN.

The centre figure may be seen bending his crossbow with a windlass, with his foot in
the stirrup of the weapon.

From Manuscript, Froissart's ' Chronicles.'

THE ordinary MILITARY CROSSBOW of the fifteenth century, with a thick steel bow, was able, if elevated to 45°, to propel its bolt from 370 to 380 yards.

The so-called point-blank range of a weapon of this description, was from 65 to 70 yards. The word point-blank must not, however, be read literally, as the head of a crossbow bolt was usually inclined a little upwards as it lay on the stock of the weapon, in order that it might have a slightly rising flight at all distances at which it was used, an arrangement in some measure applied to the sighting of our modern guns and rifles.

If one of these strong military crossbows was aimed horizontally at the forehead of a man standing at a distance of 50 yards, the bolt would not strike lower than his chin.

A few years ago I tested the shooting powers of many fine examples of crossbows, military and sporting, made near the end of the fifteenth century, some of which were formerly in a well-known Continental arsenal, and others in my own collection.

I fitted their steel bows with thick hempen strings, set their stocks and locks in order, and shot, to and fro over level ground, numbers of bolts of diverse lengths and weights, which I had caused to be made in exact imitation of the decayed originals to be found in Continental and other armouries.

The longest flight I obtained from one of the best and strongest of these weapons, originally carried by a crossbowman in battle, was 390 yards.[1] The shortest flight, from the same bow, was 380 yards. The weight of this crossbow, without its windlass, was 15½ lbs. Its steel bow was 2 ft. 7½ in. long, and at its centre 1¾ in. wide and ¾ in. thick. The former distance, in my opinion, is considerably further than any longbow archer of mediæval, or later times, could drive the arrow used in sport or in warfare.

Very few of the most powerful and skilled of modern archers, even with selected bows and light flighting arrows, are able to achieve a range of 300 yards, 280 to 290 yards being an exceptional feat. There is no reason whatever to suppose that our ancestors were so vastly superior in the use of the longbow, as to excel these distances by so much as 90 yards— especially with the heavy shafts and heads of warfare[2]—and thus to equal with the arrow of the longbow, the length of flight (370 to 380 yards) attained by the bolt of a large military crossbow with a thick steel bow.

In 'King Henry IV.,' Second Part, Act III, Scene II, Shakespeare makes Shallow exclaim of Double—'Dead! a' would have clapped i' the clout at twelve score ; and carried you a forehand shaft a fourteen and fourteen and a half.'

From this it is evident that in the time of Shakespeare, 1564-1616, it was considered a notable performance to send a forehand shaft (presumably a flight arrow) 14 to 14½ score yards (280 to 290 yards that is), which,

---

[1] 'On March 21, 1661, 400 archers, with their bows and arrows, made a splendid and glorious show in Hide Parke, with their colours flying and crossbows to guard them. Several of the archers shot near 20 score yards with their crossbows.'—Extract from *Wood's Bowman's Glory*, 1682.

[2] There is no doubt that the heads, or piles, of the war arrows used by our ancestors were far heavier than those of the target-arrows of the present day. The arrow of the ancient longbow had a barbed head, in order that its extraction might be a dangerous and difficult matter.

In the ninth year of the reign of Edward III. 'the King commanded the Mayors and Sheriffs of the county to supply 300 good bows, and four chests of arrows of the length of one ell, the heads of the said arrows to have flukes or barbs of a large size.'—*Cotton MS.*

it is curious to note, is just the range that can be reached by an unusually strong and skilful archer of the present day.

THE CROSSBOWS FOR KILLING DEER—such as the one described in Chapters XVIII–XXVIII—were somewhat lighter and less powerful than those intended for war, their bolts being of course also smaller.

I find that these sporting crossbows send their bolts at furthest 350 yards, their average length of flight being from 330 to 340 yards. The point-blank range, so called, of a good sporting crossbow, with a steel bow, was from 50 to 60 yards, which was no doubt sufficient in the days before animals had been made wary by the report of hand-guns, and when the hunter with his noiseless crossbow, could lie in wait for deer as they wandered across the glades of a forest, or visited their feeding and drinking haunts.

The extreme range of the SMALLER SPORTING CROSSBOW (Chapter XXXII), which shot a light poisoned bolt, was from 270 to 280 yards.

In the 'Dunstable Chronicle' Henry V. is described as approaching the town of Rouen 'within a distance of 40 rods or within shot of a quarrel from a crossbow.' Forty rods is 220 yards, and 'within shot of a quarrel' suggests the range of a quarrel to be further than 220 yards. This distance, however, doubtless implies merely a shot with an effective aim, and not one made to test the extreme range of a crossbow, which would certainly be far more than 220 yards when the weapon was pointed upwards at a high angle.

I have never been able to test a crossbow with a composite bow of horn, yew, and sinew, but this variety must necessarily have been much inferior in power to a crossbow with a thick steel bow. There are few weapons with composite bows in existence, and these are in such a dilapidated condition that no experiments can be made with them.

It may be taken that the ordinary longbowmen of the days of Crécy and Agincourt, could not shoot the heavy-headed war arrow to a greater range than about 250 yards.[1]

As I have pointed out, the skilled modern archer, with a flight arrow

[1] It is true the English bowmen sometimes carried several arrows with much lighter heads and shafts than the others in their sheaves. They used these to harass an enemy, and especially his horses, at a distance which was beyond the reach of the ordinary, and heavier war arrow. Though these lighter arrows probably, in some degree, resembled modern flight, or roving arrows, they must have had heavier heads than the latter to have been of any use in warfare.

Whether the English bowman of the fifteenth century could shoot his lighter arrows further than a flight arrow can be propelled by an accomplished archer of the present day, is doubtful, particularly when we consider that it is not one bow in a score that will shoot a flight arrow successfully. For instance, I have bows of 75 to 80 lbs. with which I can draw a 30 in. flight arrow to the head, but which at the same time do not drive it nearly so far as a bow with a pull of only 60 lbs.

of ordinary length, can rarely exceed a range of 280 to 290 yards, though a distance of 320 to 340 yards has on two or three occasions been recorded; Roberts, the author of 'The English Bowman,' even writes of a shot of 360 yards which he was informed occurred in 1798, but which was made, however, on declining ground.

In the days when the longbow was at its best, and was the national weapon of the English in war and sport, every man and youth, rich or poor, who could bend a bow, constantly practised archery. That some of the great number of archers continually shooting with the longbow, would be able to surpass considerably the ordinary bowman of the period in feats of range and marksmanship, was of course likely.

I am convinced, however, that none of these exceptional performers ever shot the mediæval arrow used in warfare, sport or at the target, a distance of 420 yards. I doubt if a range of even 390 yards was ever attained by an English longbowman, unless with the aid of a strong wind, or from an elevation.

Many of our castles which were built in the days when archery flourished, and before the introduction of long-range steel crossbows, are within 300 to 350 yards of eminences. The courtyard of the great castle of Carnarvon, for instance, is commanded by a hill only 330 yards distant from it.

If mediæval archers shot from 350 to 400 yards, as they are often alleged to have been able to do easily, Carnarvon Castle would never have been built where it is; as a company of bowmen could have poured their shafts into its garrison from the hill that overlooks the fortress.

Berkeley Castle is another example. The parish church at Berkeley is within 50 yards of the castle keep. Its church tower, however, stands by itself, 134 yards from the centre of the keep, and 170 yards from the courtyard of the castle. It was erected at a distance from the body of the church, in order to prevent the archers of an enemy from annoying the garrison of the castle should they happen to seize the tower as a point of vantage. There is, indeed, no other reason for the isolated position of the church tower at Berkeley.

In this case, it will be noticed that a much shorter range than that at Carnarvon was considered to be a safe one against the assaults of bowmen.

Even in modern days, the feats of shooting with the mediæval longbow are asserted to have been of such a marvellous nature, that the writers not only excite ridicule among those versed in archery, but too plainly show their ignorance of the arm.

Sir Walter Scott ('Ivanhoe') is certainly a culprit in this respect, with his accounts of the piercing of willow wands, and of an archer purposely splitting a rival's arrow when it was fixed in the target.

FIG. 12.—SHOOTING RABBITS WITH THE CROSSBOW.

From Stradanus.

[Joannes Stradanus, born at Bruges 1536, died at Florence 1605, a Flemish historical painter who delighted in portraying all kinds of sport, such as shooting, hunting, fishing and coursing, which he did with wonderful skill and in most realistic fashion. This picture is reduced from 'Venationes Ferarum,' a work consisting of 105 large plates of sporting scenes, dated 1578. Purse nets and stop nets may also be seen in use.]

The hunters carry stonebows, and the rabbits are being driven from their burrows by smoke and fire.

In his clever novel 'The White Company' Sir Conan Doyle describes a contest between the crossbow and the longbow which is simply amazing in its details, the archer finally shooting an arrow to a distance of over 600 paces![1]

The author of 'The White Company' incorrectly describes the crossbow in several details ; he even alludes to its double string.

No bolt-shooting crossbow, such as the one described in 'The White Company,' had anything but a single string. The double-stringed crossbows merely discharged stone pebbles, or else pellets of baked clay, never bolts ; the smaller kind were used by ladies and pages, and the larger by shooters of small game, such as rabbits, partridges on the ground, or, by means of a lantern, pigeons roosting in the trees at night. Yet there is a modern picture of Queen Elizabeth knocking a stag head over heels, at some hundred yards or more, with a double-stringed stonebow which, at its best, would scarce have killed a thrush at twenty paces. This is an example of how little is known of the crossbow at the present day. It is from the primitive double-stringed stone-bow of the sixteenth century that our comparatively modern, and far more powerful, rook-shooting bullet-crossbow was adapted.

The feats achieved with the longbow were proverbially enlarged upon in England as soon as the weapon became obsolete, and when the gossip of ancient archers was no doubt listened to with interest by a rising generation who could not contradict the stories they were told, and who had but slight acquaintance with the weapon. The phrase 'drawing the longbow' soon passed into a proverb, which suggested an exaggeration of the truth of any unusual performance ; yet it was, probably, pleasant enough to sit in the chimney corner of a village inn, and to listen, over tankards of ale, to the highly-coloured reminiscences of John, the archer and old soldier, or to those of Will, the tall yeoman, both of whom, maybe, had carried their bows on the fateful field of Flodden.

Whatever its extreme range may have been, there is small reason to doubt that at a distance of 150 yards the old English longbow quite equalled, if it was not indeed superior to, the flint-lock musket or 'Brown Bess' which was carried by our soldiers till about 1840.

If a hundred good marksmen armed with the 'Brown Bess' as used at Waterloo, and a hundred of the best archers of the days of Crécy and Agincourt, could be opposed to one another in line at 120 yards, the archers would, in my opinion, gain an easy victory. The archers could discharge at

[1] We read in the same chapter of *The White Company* that two other bowmen severed in eight shots the hempen cable of a large vessel moored 200 paces from the shore. Marvellous aiming this, when we consider that to cut the cable through, the eight arrows must have struck it within some quarter of an inch of each other !—and this at 200 paces !

least six shafts to every bullet fired by their opponents, and they would also, I believe, shoot with greater accuracy and effect.[1]

In connection with long-distance shooting with the bow, I append a letter written by one of my ancestors to another, who were both skilled and enthusiastic archers in their day. This letter, and the paper that follows it, describe the extraordinary distances said to have been achieved by the Turks with their bows, when shooting to attain a long range with a miniature flighting arrow.

I must explain, however, (and this goes a long way to account for the distances recorded in the letter and paper quoted,) that the flight arrows of the Turks and Persians were lighter and shorter than an English flight arrow. These Turkish and Persian arrows were only 2 ft. to 2 ft. 2 in. in length, and those which I have seen and owned, were made of bamboo. A small cap of steel or ivory acted as a head, and a little piece of hard wood as the nock, the feathering being formed of two strips of thin paper, varnished to keep it hard and upright. The arrow being so short, its head was drawn several inches inside the belly of the bow; for this reason, the forepart of the arrow was laid on a flat piece of horn about 8 in. long, with a straight groove down its centre. This horn piece was buckled in a level position along the wrist of the bow-arm of the archer, so that the arrow could be discharged without striking his wrist or the inside of the bow. In fact, the archer turned himself into a great cross-bow, and in this way he discharged a short light arrow from a very powerfu bow, and hence of course attained an immense range with it.

I need scarcely add that an arrow of this description was useless for

---

[1] As an example of what was considered a good shot with the 'Brown Bess' at about the time of Waterloo, I give the following extract.

The original MS. in which the occurrence is recorded, is in the possession of my friend—the diarist's nephew—Sir Henry Ingilby, of Ripley Castle, Yorkshire.

Extract from the Diary of Lieutenant Ingilby, R.H.A. (afterwards General Sir William Ingilby), in the Peninsular and Waterloo Campaigns.

'*May* 10, 1811.—A Spanish officer of Don Juliano's Guerillas was killed to-day through his own impru-dence. An uncommon thick fog obscured the morning, and, as the sun dissipated it, this officer made his appearance between the lines of vedettes, brandishing his sabre and making most extravagant gestures. He was as near the French vedettes as our own. Lord Wellington mistook him for a French dragoon and instantly ordered a soldier of the guard to fire at him, who, resting his musket on one of our gun-wheels, fired, and the ball passing through the head of the person, he fell dead to the ground. I witnessed myself this singular shot. The distance of it was afterwards measured and found to be 80 yards!'

NOTE ON THE 'BROWN BESS.'—The flint-lock arquebus was introduced into England from the Netherlands by William III. The last syllable of the name 'arquebus' was detached in England, and anglicised into 'Bess;' the weapon being called 'Brown Bess' either from the colour of its barrel, or of its dark walnut stock. 'Bess,' as pointed out, was an English corruption of the Dutch 'Bus,' a barrel. The word 'Bus' was formerly applied indifferently to a barrel or a gun. For example, 'Donner-bus' in Dutch meant the 'thundering barrel,' but was changed in England to blunderbuss.

Again, 'Handbus' was a pistol, literally a hand-gun, and 'Bus-schieter' a gunner or barrel-shooter. The transition of the name from Arquebus to Brown Bess may be taken as :—The Brown Arquebus—Brown Bus (*i.e.* brown barrel or gun)—Brown Bess.

warfare, or even for target-shooting, as it would break to pieces on striking any material that was more resistant than sand or soil.

'London 1795.

'Dear Brother,—I have just been to see the secretary of the Turkish Ambassador shooting with Waring[1] and other famous English bowmen. There was a great crowd, as you may suppose, to see them. The Turk, regardless of the many persons standing round him and to the amazement and terror of the Toxophilites, suddenly began firing his arrows up in all directions, but the astonishment of the company was increased by finding the arrows were not made to fly, but fell harmlessly within a few yards. These arrows the Turk called his " exercising arrows." This was an idea that was quite new to the bowmen present, and they began to have more respect for the Turk and his bow. The Turk's bow is made of antelopes' horns and is short, and purposely made short for the convenience of being used in all directions on horseback.

'The Toxophilites wished to see the powers of the Turkish bow, and the Turk was asked to shoot one of his flight arrows. He shot four or five, and the best flight was very carefully measured at the time. It was 482 yards. The Toxophilites were astonished, I can tell you.

'Waring said the furthest distance attained with an English flight arrow, of which he had ever heard, was 335 yards, and that Lord Aylesford had once shot one, with a slight wind in his favour, 330 yards. Waring told me that he himself, in all his life, had never been able to send a flight arrow above 283 yards.

'The Turk was not satisfied with his performance, but declared that he and his bow were stiff and out of condition, and that with some practice he could shoot much further than he had just done.

'He said, however, that he never was a first-class bowman even when in his best practice, but that the present Grand Seigneur was very fond of the exercise and a very strong man, there being only two men in the whole Turkish army who could shoot an arrow as far as he could.

'The Turk said he had seen the Grand Seigneur send a flight arrow 800 yards.

'I asked Waring to what he attributed the Turk's great superiority over our English bowmen; whether to his bow or not. Waring replied he did not consider it was so much the result of the Turk's bow, but rather of his strength and skill, combined with the short light arrows he used, and his method of shooting them along the grooved horn attached to his arm.

'Neither Waring nor any of the Toxophilites present, (and many tried,) could bend the bow as the Turk did when he used it.

---

[1] T. Waring, author of a *Treatise on Archery*, 1st ed. 1814, last ed. 1832. Waring was an accomplished archer and a well-known manufacturer of bows and arrows.

'So much for the triumph of the Infidels and the humiliation of Christendom.
'Yours aff.,
'W. FRANKLAND.

'To Sir Thos. Frankland, Bt., M.P.
'Thirkleby Park.'

I found the following in a manuscript notebook of 1798 describing feats and incidents of archery, collected by the recipient of the above letter.

———————

'Records of Turkish archery procured in 1797 from Constantinople by Sir Robert Ainslie, at the request of Sir Joseph Banks, and translated by Sir Robert Ainslie's interpreter.'

'The Turks still have detachments of archers in their armies, merely not to deviate from ancient custom, for, in Turkey, archery is now merely regarded as an amusing exercise that is to this day practised by all ranks of the people.

The Ottoman emperors, with their court, often enjoy the diversion of archery in public, and there is an extensive piece of ground allotted to that purpose.

This place is upon an eminence in the suburbs of the city of Constantinople, and commands an extensive view of the town and harbour. It is called Ok Meydan, or the Place of the Arrow. The ground mentioned is covered with marble pillars erected in honour of those archers who have succeeded in shooting arrows to any remarkable distance. Each pillar is inscribed with the name of the person whose dexterity it records, together with some complimentary verses to him, and the exact range which he attained with his flight arrow.

The Ottoman emperors, from ancient times, have been always supposed to live by their manual labour, and in consequence of this supposition they have each learnt some art or profession, most of them having preferred the art of making bows and arrows.

The present emperor was bound apprentice to the trade of archery, and at the time he was received as a master in this trade, he gave on different occasions very splendid public entertainments at the Ok Meydan, where the State tents were pitched for him and his court.

The Tartar bows are preferable to those manufactured in Turkey, as the former are the larger and stronger, though there is now an extensive factory for implements of archery in Constantinople, called Ok Zilar, or the place of the Arrow-makers.

The Turkish bow is formed of a very strong elastic wood. One side of the bow is covered with a composition made chiefly of buffalo horn melted down; this is smoothed with a file to a proper shape, and forms the concave side of the bow when it is bent.

The convex side is plain wood, painted, varnished and richly gilt. The bow is only bent when it is about to be used, and then it is bent with much caution, the heat of fire being always first employed to make it flexible.

The Turkish bow will penetrate, with an ordinary arrow, a half-inch plank at over 100 yards, the head and shaft of the arrow passing for three or four inches through the wood.

Translations of the inscriptions on some of the marble columns at the Ok Meydan (Place of the Arrow), which were erected in honour of those who have excelled in archery.

1. Ak Siraly Mustapha Aga shot two arrows both of which travelled to a distance of . . . . . . 625 yards.
2. Omer Aga shot an arrow to a distance of . . . . 628 ,,
3. Seid Muhammed Effendy, son-in-law of Sherbetzy Zade . 630 ..
4. Sultan Murad . . . . 685 ,,
5. Hagy Muhammed Aga shot an arrow . . . . 729 ,,
6. Muhammed Ashur Effendy shot an arrow which fixed in the ground at . . . . 759 ,,
7. Ahmed Aga, a gentleman of the Seraglio under Sultan Suleiman the Legislator, shot an arrow . . . 760 ,,
8. Pashaw Oglee Mehmed shot an arrow . . . . 762 ,,
9. The present Grand Admiral Husseir Pashaw shot an arrow which drove into the ground at . . . . . 764 ,,
10. Pilad Aga, Treasurer to Hallib Pashaw . . . . 805 ,,
11. Hallib Aga . . . . 810 ..
12. The reigning Emperor Sultan Selim shot an arrow which drove into the ground at a distance of . . . . 838 .,
   The Sultan shot a second arrow to near the same distance.'

In the translation of the above from the Turkish language, the feet and inches are also given for each shot, but these I have omitted as unnecessary.

In the manuscript, the interpreter remarks that the measurements of the distances on the marble columns at Ok Meydan are in pikes, the pike being a Turkish measure of a little over two feet, easily convertible into English yards, feet and inches.

It will be observed that the longest flight recorded on the columns selected for quotation is 838 yards, and the shortest, 625 yards. Though these distances are almost too extraordinary to be true, they corroborate in some measure the statement made in 1795 by the secretary of the Turkish ambassador, p. 27.

If they are correct, they can only be accounted for by the use of a light short arrow, a very powerful bow, great strength and skill, and, above all else, by the horn appendage which the Turkish archer attached to his left arm, and without which he could not shoot so short an arrow from his bow.[1]

If a very light flight arrow of reed or bamboo could in some way be arranged to receive the impulse of the thick string of a crossbow with a powerful steel bow, I have little doubt it could be propelled half a mile.

I have fitted (as a separate piece) a large hollow horn nock over the butt of the ordinary flight arrow of the longbow, so that the loose nock rested against the string of the crossbow. In this way I have obtained several flights of from 500 yards to 515 yards. In the case of a short and very light flighting arrow, however, the recoil of the steel bow shivers it to pieces as it leaves the stock of the crossbow.

[1] Even if we accept only the shortest range recorded on the columns as correct—*i.e.* 625 yards—it is an extraordinary distance for any arrow to be propelled, and much exceeds, as far as we know, what has ever been done by an English bowman with a longbow. It is, however, beyond question that the secretary to the Turkish Ambassador did shoot an arrow 482 yards (the arrow and bow being even now preserved in the Toxophilite Society's rooms), though he declared at the time of the occurrence that he was not proficient in the art of sending a flight arrow to what he considered a great distance. We may from this safely assume that a range of 143 yards further than the Turkish secretary attained with his bow, or a total flight of 625 yards, was quite possible in the case of a more powerful and skilled Turkish archer than he was.

See Chapter L. for a description of long distance arrow-throwing by hand.

CHAPTER VI

THE SHORTBOW AND LONGBOW IN RELATION TO THE CROSSBOW

FIG. 13.—ARCHER AND CROSSBOWMAN
OF ABOUT 1370.

The kneeling figure is fitting his belt-claw to
the string of his crossbow, preparatory
to bending its bow.[2]

*From Manuscript No. 2813 in the National Library,
Paris, reproduced by J. Quicherat in his
'History of Costume in France,' 1875.*

IN the Bayeux Tapestry, though no crossbows are shown, many Norman soldiers are depicted carrying bows and arrows, the bows being shortbows and not longbows. The longbow can usually be identified, as its length was about the same as the height of the man who carried it. In this pictorial and contemporary representation of the Conquest of England, there is only one British bowman to be found, and he is bearing the ordinary Saxon shortbow.[1] The bow was little used by the Saxons at the time of the Conquest, their chief weapons at that time being spears and axes, both of which they cast at the enemy when he approached sufficiently near.

The shortbow, and the primitive crossbow with its bow of solid wood, or of wood, horn and sinew, were probably equally effective in early mediæval warfare, the crossbow being, perhaps, the more efficient of the two weapons in the case of men wearing mail, or carrying leathern shields.

It was when the powerful longbow, as used by foot-soldiers only, appeared, that the crossbow with its wood, or horn and wood, bow was completely overmatched.

---

[1] See Plate LXV. in the history and description of the Bayeux Tapestry by F. R. Fowke, 1898.

In reference to the above, Edward A. Freeman in his *History of the Norman Conquest of England*, vol. iii. p. 472, writes ' Only one English archer is represented in the Bayeux Tapestry.'

[2] See Chapter XV. for a description of the belt-claw.

The longbow held its own, as by far the most deadly manual missive weapon in warfare, till about 1370, when the crossbow with its thick steel bow and powerful windlass was introduced.

The shortbow, the ancient form of bow, though carried by foot-soldiers in early mediæval times, was more often the arm of mounted men, especially abroad. Being short, it could conveniently be discharged from horseback, and when not in use it could be slung over the back of the soldier with its cord across his breast. The string of the shortbow was drawn to the breast, and not to the right ear as in the longbow. The longbow could not be used on horseback, but in the hands of footmen it was an infinitely more powerful arm than its predecessor, the shortbow.

FIG. 14.—SHOOTING AT THE BUTTS WITH CROSSBOWS.
From MS. Royal Library, dated 1496, reproduced by J. Strutt in his 'Sports and Pastimes of the People of England,' 1801.

With the English, the longbow, which was never popular abroad, gradually usurped the place of the shortbow. In the assize of arms fixed by Henry II. in 1181, bows, whether short or long, are not even alluded to as weapons of the period. It was only in the last quarter of the thirteenth century, 1272–1300, during the reign of Edward I., that with English troops the longbow became a popular weapon, and in great measure superseded the shortbow and the primitive crossbow. Subsequent to about the year 1340, English soldiers carried longbows only, and never, or very seldom, crossbows. After the successes of the English longbowmen at Falkirk in 1298 —the first notable triumph recorded of these weapons—and especially after the splendid victory gained by their assistance at Crécy in 1346, and again at Poitiers in 1356, and at Agincourt in 1415, our ancestors naturally despised the crossbow as a military weapon.[1]

---

[1] The contemporary French chroniclers of the battle of Crécy, allude to the English longbow as being at that time a new and deadly weapon in Continental warfare.

Many of the mercenary troops, however, such as Genoese and Gascons, whom we constantly hired in mediæval days to fight for us abroad, and occasionally at home, were armed with crossbows till about 1480.

Crossbows for killing deer, and for shooting at butts, were fairly common among the English in the fifteenth century, and it was doubtless recognised by those in authority, that if the people practised with these easily manipulated weapons instead of with their longbows, skill in the use of the latter might be wanting in time of national danger.

It was, therefore, with certain reservations, as in the case of nobles and persons of wealth, at length enacted, that the possession of a crossbow, even for sporting purposes, be forbidden by law among the people of England.

This Act was introduced in 1508, during the reign of Henry VII., and reinforced by statute, mandate, or proclamation, in 1512, 1515, 1524, 1528, and 1534 in that of his successor Henry VIII. In 1536, the Act against crossbows was repealed, and their use was permitted, except in the King's Parks and Forests.

In 1537, the Act was once more renewed; this time it included hand-

FIG. 15.—CROSSBOWMEN PRACTISING AT THE TARGET.

Their dogs are retrieving the arrows, and were trained to do this without injuring the feathers of the missiles.

From a translation into Italian of 'A History of the Peoples of the North,' by Olaus Magnus, Archbishop of Upsala. Printed at Venice, 1565.

guns as well as crossbows, with the proviso, that those persons who were permitted to carry hand-guns must have none that exceeded two and a half feet in length, including the stock. In the few licenses granted to various persons, such as foresters and keepers, to carry crossbows to kill game, the heron was always excepted, as heron-hawking was the favourite sport of royal and noble falconers.

The prohibition of the crossbow and the hand-gun must have been rigidly enforced in England, at all events till 1539. In April of that year, John Marshall writes to Thomas Cromwell, Lord Great Chamberlain (in 1539 made Earl of Essex, and the following year beheaded on Tower Hill): 'Have had the King's orders to provide four men to send to my Lord Admiral upon an hour's warning. Have done so. There are no gunners here by reason of the statute against crossbows and hand-guns.' (State Papers, Reign of Henry VIII.)

In 1542, the last statute against crossbows and hand-guns was passed by Parliament. This one imposed the very heavy fine of 20*l.* on anyone keeping a crossbow, and stated among other reasons for the suppression of the weapon 'that divers murders had been perpetrated by means of crossbows, and that malicious and evil-minded people carried them ready bent and charged with bolts, to the great annoyance and risk of passengers on the highways.'

The prohibition of the crossbow in England was not, it will be understood, the result of a fear that, as a superior arm, it might usurp the position of the longbow, for when the first three statutes were passed to suppress it (1508, 1512, 1515) the crossbow had been almost supplanted by the hand-gun in Continental armies, and at the dates of the later Acts, (1537, 1542) it was unknown in warfare. All the statutes against crossbows and hand-guns were introduced to prevent the yeomen and peasantry of England from practising with, or even handling a weapon of any kind other than the cherished longbow, though the later statutes may have been suggested by a fear that the hand-gun might cause the people to put less trust in the longbow than formerly, and thus in some measure to discontinue its use.

The great victories achieved with the English longbow in former days, induced English kings, and commanders of troops, to believe that no weapon ever invented or likely to be invented, whether crossbow or hand-gun, could compete with it. For this reason, the longbow was retained in English armies beyond the days of its real effectiveness in warfare, though even then, its decadence was not due to its inferiority to the hand-guns of the period, but to a scarcity of archers trained to its proper use.

Even when it was realised (1570–1580)[1] that the longbow was being hopelessly beaten by the hand-gun in battles and sieges, and had no chance of regaining its position, several statutes were passed, all of course unavailing, with a view to saving it from extinction as our national and well-tried weapon.

The longbow was at its best from the time of Crécy, 1346, to about 1530. It began to decline in favour about 1540.

In the large engraving of the picture of the siege of Boulogne in 1544, and in the one of the fight between the English and French fleets off Portsmouth in 1545 (the original pictures were burnt at Cowdray House, in 1793), there are as many English soldiers depicted with hand-guns and pikes as with longbows.

In Latimer's sixth sermon, printed in 1549, the Bishop bewails the decline of the English longbow, and calls upon the magistrates of England to do their

_____
[1] At this time the longbow was, however, quite as effective as any hand-gun. Its decadence was due to a neglect to practise with it during the more or less peaceful reign of Elizabeth. See p. 39, for Montaigne's criticism of hand-guns at this period.

duty, and enforce the statutes that direct the peasantry to possess longbows, and frequently to practise with them.

On the other hand, foreign nations hailed with delight the gradual disappearance of the English longbow, which they had had such good cause to dread for so long a period of history. Hence our Continental enemies encouraged the use of the hand-gun, as an arm which might place their soldiers, whether young or old, on an equality with the tall and strong English archers, and which, unlike the longbow, required no special strength to manipulate.

After a persistent struggle against gunpowder, the longbow was generally discarded by the English between the years 1580 and 1590. It was employed in desultory fashion till about 1615, on a few occasions as late as 1620-1630 (notably in the expedition to the Island of Rhé in 1627), and still more recently, and for the last time in regular warfare in our islands, by Montrose, in his defeat of the Covenanters at Tippermuir, near Perth, in 1644. The longbow was, however, used in the Highlands of Scotland in tribal disputes at a later date than 1644, or long after it was laid aside in England and the Lowlands.[1]

For instance, in September 1665 the Mackintosh gathered his clan and entered Lochaber, the territory of Cameron of Lochiel, these two Chieftains having been at feud for many years.

Lochiel, of course, assembled his adherents to repel the invaders, and found, on taking muster, that he had at his disposal 900 men armed with guns, broadswords and shields, besides 300 more men who carried bows instead of guns. ('Memoirs of Sir Ewen Cameron of Locheill;' printed at Edinburgh, 1842.) The compiler of the Memoirs remarks, 'This was the last considerable company of bowmen seen in the Highlands.'

The historic reputation of the longbow was so great in England that several pamphlets were issued during the eighteenth century, advocating its re-introduction as a military weapon.

Even so recently as 1798 a book was published with this title:

'PRO ARIS ET FOCIS.

'CONSIDERATIONS OF THE REASONS THAT EXIST FOR REVIVING THE USE OF THE
LONGBOW WITH THE PIKE[2] IN AID OF THE MEASURES BROUGHT FORWARD
BY HIS MAJESTY'S MINISTERS FOR THE DEFENCE OF THE COUNTRY. BY
RICHARD OSWALD MASON, ESQUIRE, LONDON 1798.'

[1] This is confirmed by innumerable passages in the criminal records, and in the record of the Privy Council, of Scotland. The last time archers employed the longbow in warfare in Scotland, is said to have been at a great clan-battle, fought in 1688, between the Laird of Macintosh and Macdonald of Keppoch.— *Archæologia Scotica*, vol. iii.

[2] In this case the pike was to be employed as a separate arm, and not as recommended by William Neade in his curious book *The Double-Armed Man*, published in 1625. In Neade's book, the bow was attached to the pike when the latter was used to repulse cavalry.

The fact that the hand-gun could be used by the horse-soldier, whilst the longbow could not be thus employed, was the chief argument against the revival of the bow.

There is no doubt, that our Continental enemies, after their experience of the longbow at Crécy in 1346, held it to be a fearfully destructive weapon. They, in fact, had little knowledge before that battle of its greatly superior power and accuracy in open warfare to all other missive arms of the period, such as, for instance, the crossbow with a composite, or a light steel, bow.

Though the French sometimes carried the longbow in the chase, they never succeeded in mastering it as a weapon of war, despite strenuous efforts to do so. After Crécy, the French endeavoured to introduce it into their

FIG. 16.—MOUNTED CROSSBOWMAN.
From 'Famous Women,' by G. Boccaccio. First Edition, 1473.

armies, with a view to combating the English with their own weapon, any soldier who excelled in its use being highly rewarded. The French, however, came to the conclusion that they could never handle the longbow as did the English, and, for this reason, they returned to the crossbow as their favourite arm. Père Daniel writes, 'The French King did manifestly see that neither his nor any other people could attain to shoot so strong, and with that dexterity and excellence which the English bowmen did, whereby, and seeing that English archery was a very peculiar gift of God, they left off the practice and use of the longbow.' [1]

At the time of Crécy, the armour worn by the knights was not designed

[1] Père Daniel, Superior of the Jesuits at Paris, French historian, born 1649, died 1728.

to withstand the powerful longbow, and suits of chain mail, or of light plate, were almost useless as a defence against its arrow. It was only when chain mail was discarded and plate armour was made heavier, that both the arrow of the longbow, and the bolt of the crossbow which had a thick steel bow, became less dangerous to knights and foot-soldiers. A shower of two or three thousand arrows falling from aloft must have been a terrifying sight, especially to a body of cavalry standing or moving in close rank. Bullets from the primitive hand-gun had a comparatively low trajectory and short range, and could not be detected as they passed through the air. On the other hand, every soldier could see a cloud of arrows approaching him, and he would surely imagine that one of the great number descending must strike him.

Horses, too, were driven frantic by the English bowmen, so we read, for their arrows caused the animals to rear and plunge and gallop madly in all directions, thus throwing into dire confusion any formation they were in. A bullet from a hand-gun might strike a horse, and cause him to kick or plunge only at the moment of contact, but a barbed arrow sticking deep in his flesh would, with every movement of the animal, gall and fret him beyond the control of his rider, who would probably soon be unhorsed, to become, if in heavy armour, an encumbrance on the field for the remainder of the battle.

Various writers on archery and mediæval warfare, have asserted that the longbowman was able to discharge ten to twelve arrows in the time taken by the crossbowman to shoot off one bolt. But the crossbow was not nearly so slow as alleged, and experiments I have made to test the question of its speed in shooting prove this. A military crossbow of the fifteenth century, 15 lb. in weight, can be discharged at a mark once in a minute. The operation includes (1) Taking the weapon from the shoulder. (2) Unhooking a windlass from a waist-belt. (3) Fitting the windlass [1] to the stock and string. (4) Winding up the bow. (5) Arranging the bolt and, after taking aim, pressing the trigger. I find that a longbow can be discharged six times at a target in the space of one minute. The operation in this case also includes a fair aim, besides taking the arrows from the ground, fitting them to the string, and drawing and releasing the bow. I do not, however, imagine that either the longbowman, or his rival with the crossbow, often used their weapons in warfare with great rapidity, or their sheaves of arrows and bolts would soon have been exhausted.

---

[1] The cranequin, or ratchet-winder, Chapters XXX., XXXI., though rather slower to use than a windlass, was, however, far more convenient to manipulate, and also enabled a much smaller stock to be fitted to a crossbow than was possible with a windlass and its cords. For these reasons it was chiefly carried by mounted soldiers and by hunters of deer.

The cranequin was introduced at a considerably later date than the windlass, see p. 134.

CHAPTER VII

THE HAND-GUN IN RELATION TO THE CROSSBOW

THE reliance placed by English commanders on longbows, and by Continental captains on crossbows, made the introduction of hand-guns a very slow process. The French and Spaniards were the last nations to discard the crossbow for the hand-gun, the French being particularly averse to the latter weapon, though they seem to have employed fire-arms (cannon) for many years previously to arming their soldiery with hand-guns.

Hand-guns must not be confused with fire-arms, as cannon were invented long before hand-guns, and are even said to have been used as early as 1346, at Crécy. Though Gibbon, the historian, doubts the presence of cannon at Crécy, and Froissart does not mention them, yet Villani,[1] who died probably within two or three years after the battle, and later De Montluc,[2] positively assert that the cannon brought to Crécy by Edward III. materially assisted in the victory of the English.

The hand-gun became more or less popular, ineffective as it was, with various Continental nations and States many years before it was used in France. For instance, hand-guns were employed by the Hussites in their revolutionary wars in Bohemia (1419-1436), as well as by the Florentines at the siege of Lucca in 1431.

According to Sismondi,[3] the Milanese armed their militia in 1449 with these new weapons, and at the battle of Morat in 1476, when the Duke of Burgundy was so signally defeated by the Swiss, the victors had among their troops 6,000 men who carried hand-guns.

At all events, the hand-gun took the place of the crossbow on the Continent at a much earlier date than it superseded the longbow in England.

---

[1] G. Villani—Florentine historian, wrote *Storie Fiorentine*, born about 1280, died about 1348. The *Storie* were continued to 1363 by his brother Matteo, and then to 1364 by Matteo's son Filippo.

[2] De Montluc. Blaise de Lasseran Massencomé, Seigneur de Montluc, born 1503, died 1577, Marshal of France. Wrote the memoirs of his career as a soldier, which were termed by Henry IV. of France the *Soldier's Breviary*. Was made a captain under Francis I. in 1523. His life is given in Petitot's *History of France.*

[3] S. de Sismondi—Swiss historian, born 1773, died 1842.

The Seigneur de Montluc,[1] who fought so gallantly for Francis I. in his wars with Charles V. of Spain, has left on record in his Commentaries, which so ably describe his fifty years of active service, 'that when he first commanded troops (1518–1520) under Francis I. only crossbowmen were in the French army, and not one soldier with a hand-gun.' It is, however, recorded that at the siege and capture of Turin in 1536, hand-guns had quite superseded crossbows, and that only one crossbowman was then present in the French army, though this man was so clever with his weapon that he killed therewith more of the enemy than were killed by the best hand-gunner present at the siege.[2]

The first hand-guns seen in England, were carried by the Burgundian troops under Warwick, at the second battle of St. Albans in 1461. In 1471, when Edward IV. landed at Ravenspur, a port then existing on the north shore of the Humber close to its entrance to the sea, he brought among his troops 300 Flemings armed with hand-guns.

It is difficult to understand the increasing popularity abroad of the miserably ineffective hand-gun, unless it was persistently encouraged as a rival to the English longbow.

Throughout the greater part of the sixteenth century, foot-soldiers with hand-guns, without the support of cavalry, would have been an easy prey in open field of battle to men armed with longbows, who were properly trained to use them.

In 1585, Montaigne [3] wrote 'that the effect of the discharge of a hand-gun, apart from the shock caused by its report, was so insignificant that he hoped the use of these weapons in warfare would soon be discontinued.'

Another chronicler records that at the battle of Kissingen in 1636, the slowest soldiers fired only seven shots with their hand-guns during eight hours, and that at Wittenmergen in 1638, the soldiers of the Duke of Weimar fired off their pieces only seven times each man, and this, too, during an engagement which commenced at noon and lasted till nightfall![4]

[1] See Note 2, p. 38.

[2] *Discipline Militaire.*—Doubtfully attributed to Guillaume de Bellay, French general and historian, born 1491, died 1543.

[3] Michel de Montaigne—French moralist and author of *Essais*, born 1533, died 1592.

[4] Even a century after Wittenmergen the musket was a very inferior arm, and the powder of its time so weak that an immense charge was used. In 'Art de la Guerre, by the Marquis de Puységur, Marshal of France, printed 1748,' the author writes, 'We lose some men at 200 paces, more at 100, and still more at 50 paces.' In 'Tactical Training of the Prussian Army, 1745–1756, by Frederick the Great,' we read of his infantry musket, that its calibre = 20·14 mm., bullet = 31·3 grammes, charge of powder = 19·53 grammes and that though fire was opened with it at 300 paces, it only became effective at 200 paces (i.e. 167 yds.). In a trial of the Prussian musket, about 1810, only 50 out of each 100 bullets that struck it, pierced a pine-wood target one inch thick, at 200 paces.

CHAPTER VIII

SUMMARY OF THE DEVELOPMENT OF THE MEDIÆVAL HAND-GUN

In the heading to this chapter, I have used the word 'hand-gun' to express any hand fire-arm that was carried by the individual soldier in mediæval times.

Hand-guns were first seen in warfare at the end of the fourteenth century, and were then known as hand-cannon. They were merely small reproductions of the fire-arms or cannon which for many years previously had been employed in sieges.

The earliest hand-gun consisted of a short metal tube, of $\frac{1}{2}$ in. to $\frac{3}{4}$ in. bore, with a touch-hole on the top of its breech-end, like a cannon.

This tube was fastened to a straight piece of wood, either by means of small iron hoops, or by thongs of leather.

The weapon was discharged by placing a burning fuse to the priming powder which was piled up over its touch-hole. The first hand-gun was, in fact, a miniature cannon, made light enough to be manipulated by one man, and with a handle fastened to its breech-end by which to hold and direct it. This form of hand-gun was in limited use in foreign armies till about 1460.

For a long time after their introduction, the smaller hand-guns had straight narrow stocks, similar in shape to those of military crossbows, the pointed end of the stock of the hand-gun, as in the larger crossbow, being rested upon the top of the right shoulder when aim was taken, fig. 4, p. 8.

Even the manner of sighting over the thumb, as it lay on the top of the stock, was also copied from the crossbow, a primitive system of alignment retained in the hand-gun for many years.

The straight crossbow-shaped stock was not generally discarded in hand-guns, and the enlarged butt-end for the shoulder substituted, till about 1500.

The next variety of hand-gun was very heavy, and was known as a culverin. Small culverins were, however, carried by horsemen, but the larger kind, which weighed from 16 to 18 lbs. and was used by the foot-soldier, required the assistance of an attendant to work it.

The butt-end of the stock of the culverin was sloped downwards like the handle of a pistol.

The culverin was at length made sufficiently portable to be worked by the soldier himself, and was then known as an arquebus. It was further improved by having its touch-hole bored at one side of its barrel instead of on the top, and it was also fitted with a projecting flash-pan, placed level with the touch-hole. This pan held more conveniently the very liberal pinch of priming powder which, on being flashed by means of a burning fuse applied by hand, ignited the main charge inside the barrel. At the battle of Morat in 1476, the Swiss had 6,000 men armed with these weapons.

Both the large culverin and the smaller one known as an arquebus, were aimed and fired with their barrels resting on a forked stick.

Between 1510 and 1520 the arquebus above described was superseded by the match-lock arquebus. In this kind of hand-gun the first attempt at automatic ignition appears. It had a long hammer, pivoted in the stock, which held a piece of slow-match in its jaws. The hammer was continued in one piece through the stock, and projected beneath it in the form of a trigger.

When the trigger was pulled back, its upper half, or hammer-end, pressed down the burning fuse it held, till the fuse touched the priming powder in the pan, and thus fired (or did not, according to the weather and other conditions) the weapon. It is probable that during a battle, the fuse held by the hammer of the match-lock had to be continually rekindled by the soldier to keep it in serviceable order.

The match-lock arquebus is even now to be seen in the hands of some remote tribes of northern India.

The wheel-lock was invented 1550–1560. This hand-gun had a small wheel, about the size of a crown piece, and $\frac{1}{4}$ inch thick, which revolved at the side of its barrel. The upper edge of this wheel slightly projected, from underneath, through the centre of the flash-pan, near the touch-hole. The wheel had a serrated edge like a coarse file, and on one side of it there was a strong circular spring, and a catch to secure it when it was wound up.

The hammer of the wheel-lock gun held a piece of detonating composition or pyrite, which on being rubbed against rough metal emitted sparks

like a flint.   When the wheel-lock was in use its hammer was not raised, but—
by means of a spring which kept it down—was always pressing the pyrite it held,
hard against the top of the serrated wheel, where the latter projected through
the flash-pan near the touch-hole.

To prepare the wheel-lock for firing, its wheel was turned round a couple
of times with a key, till it was secured by the small catch which prevented
its spring from unwinding it.   The priming powder was next dropped into the
flash-pan.

On pulling the trigger, the wheel was set free, and rapidly revolved.
The serrated edge of the wheel, grinding hard against the composition held
in the jaws of the hammer, created the sparks which fired the priming, and the
charge inside the barrel.

The fire-lock, or flint-lock, was produced in Spain about 1625, and was at
first a mere adaptation of the wheel-lock, a piece of flint being fixed in the end
of the hammer instead of the pyrite.   After a short time, the revolving wheel
was discarded, and the inside of the flash-pan was made rough, so as to cause
the edge of the flint to break against it on the fall of the hammer, and to
emit sparks.

Unlike the wheel-lock ; the hammer was now cocked and released every time
the weapon was discharged, the springs of the lock acting as in a modern gun.

It was not till 1670–1680, that the flint-lock was sufficiently improved
to become generally adopted in war and in the chase.[1]

For some years, the priming in the flash-pan of a flint-lock was exposed
to wind and wet, though finally a snap cover was invented to shield the
pan and keep the priming dry.   The flint was now made to strike against
the roughened surface of the inside face of this hinged cover.   As the flint
struck the cover it knocked it back, and at the same time exposed the flash-pan
underneath it, and projected the sparks caused by the friction of contact into the
priming.

Flint-locks were introduced into England 1690–1700, and, with slight
modifications, were carried by our soldiers till 1840.

[1] At the battle of Dunbar in 1650, the Cromwellian musketeers carried match-locks, in preference to
wheel-locks or flint-locks.

It is, however, recorded that they could not use their weapons with full effect, as the heavy rain prevented
the fuses from being kept alight.

CHAPTER IX

A SUMMARY OF THE HISTORY OF THE CROSSBOW

Fig. 17.—Crossbowmen killing Deer and Wild Boars.
From MS. Gaston Phœbus. Fourteenth century.

THE Romans employed a large machine on wheels that was wound up by a windlass turned by several men, and which was made on the same principle as a crossbow. They also appear to have used the ordinary small crossbow carried by hand, even so long ago as the fourth century. Good evidence of this is to be found in Vegetius.[1] This author, in his treatise on military art, dedicated to Valentinian II. about 385, alludes to the crossbow as being a manual weapon assigned to light-armed troops, the description of which he omits, as it is so well known. Two Roman bas-reliefs, evidently older than the fourth century, described in 1831 by M. Aymard, and belonging to the

---

[1] 'Erant tragularii, qui ad manuballistas vel arcuballistas dirigebant sagittas,' Book II., Chapter 15,—with them were the javelin men, who from their bows in hand or crossbows directed their arrows.

museum at Puy, in France, present, writes Victor Gay,[1] all the characteristics of the primitive crossbow carried by hand.

From the fifth to the tenth century, all evidence, historical or pictorial, is wanting as to whether the crossbow was in common use or not.

In the tenth century, the crossbow was, however, a popular arm, as the two following extracts quoted by Victor Gay prove. These are taken from the printed reproduction of the tenth century MS. of the monk Richer's 'Historia.'

1. King Lewis, with an army from Belgium, enters the territory of the Duke, in 947. First he attacks the city of Senlis . . . on both sides very many are wounded. But the Belgians, because they were attacked vigorously by the crossbowmen of the city, could not resist . . . so by order of the king they draw off from that city, not only by reason of the crossbowmen, but also because of the strength of the towers.

2. And here Lothair, with 10,000 men, made for Verdun, 985. The bowmen were set against the foe, and the arrows discharged, and the missiles from the crossbows, were careering about in the air so thickly, that they seemed to be coming down from heaven and rising out of the earth.

In the picture of the martyrdom of St. Sebastian, which took place in 288, and was a favourite subject for Italian masters in the fifteenth and sixteenth centuries—such as Mantegna, Veronese, Domenico—the Saint is often shown as being pierced by his assailants with bolts from crossbows. The crossbows are in some cases minutely depicted, as well as the manner of stretching their strings, especially in the large picture by Pollajuolo[2] dated 1475, now in the National Gallery in London, fig. 31, p. 74.

It is curious that the crossbow is not to be found in any illustrated manuscript of the time, as having been used at the landing of the Normans in England in 1066, at the battle of Hastings, or during the subsequent conquest of England, 1066–1071. The Bayeux tapestry[3] with its hundreds of knights and soldiers, with swords, spears, bows and arrows, is contemporary with, and was specially designed to commemorate the Norman invasion of Britain, and the attendant battles, yet it does not contain one crossbow. We know, however, that crossbows were brought to England by the Normans in 1066, for they are distinctly alluded to in a contemporary

[1] Victor Gay, *Archæological Glossary of the Middle Ages and the Renaissance*. Paris, 1887.
[2] Antonio Pollajuolo, Italian painter, born 1429, died 1498.
[3] This piece of work is 230 feet long by 20 inches wide. It contains 623 persons, besides 762 horses, dogs and other animals, 27 buildings, and 41 ships and boats.

poem by Guy of Amiens,[1] as having been used at Hastings. William of Poitou[2] also writes that crossbows were carried by the Norman soldiers at Hastings. Again, Sir S. R. Meyrick, in his great work on ancient armour, states 'that in Domesday Book (1085–1086) the name of "Odo the crossbowman" is given as being a tenant of some lands of the king in Yorkshire.' The name 'Odo' shows that this man was a Norman.[3]

The historian, Thorne,[4] writes 'that at the battle of Hastings the Normans entered the field with "drawn" bows (arcubus tensis).'

This applies, I consider, to crossbows, and intimates that the Normans were prepared for an immediate assault on their opponents.

There would be no occasion for the chronicler to record as worthy of comment, that the ordinary bows of the Normans were 'strung,' as every bow would be strung, as a matter of course, some time before the battle commenced. It would, indeed, be remarkable if they were carried into action un-strung.

Besides this, the word 'tensis' suggests a bow with its string stretched, as a crossbow that was kept ready for use at a moment's notice.[5]

The string of a crossbow could be retained in this position, or in a state of tension, as when drawn back over its catch.

On the other hand, the string of an ordinary bow could not be thus held, as with this weapon, when the string was drawn it was instantly released.

William II. was accidentally (so 'tis said) killed in 1100, when hunting deer in the New Forest, by a bolt from the crossbow of Sir Walter Tyrrel. This again proves that the Norman invaders brought crossbows to England, and that they carried them for purposes of sport as well as for use in warfare.

William II., Henry I., Stephen, and Henry II. all employed crossbowmen, chiefly foreign mercenaries, in their armies.

---

[1] Guy, Bishop of Amiens. Wrote *Carmen de Hastingæ proelio*. He completed this famous poem about 1068, and died about 1075.

[2] William of Poitou, Archdeacon of Lisieux, chaplain and biographer of William the Conqueror, flourished about 1087, was born about 1020, died about 1090. Untrustworthy as a chronicler owing to his tendency to magnify the deeds of his patron, but probably correct in such a small matter as the use of crossbows at Hastings.

[3] E. A. Freeman, in *The History of the Norman Conquest of England*, Vol. iii., p. 467, writes, 'First in each division marched the Archers, Slingers, and Crossbowmen.'

[4] William Thorne, flourished about 1397, a monk of St. Augustine's, Canterbury. He deals with the general history of England. Much of his writing is derived from the chronicles credited to Thomas Sprott (flourished about 1265), who was also a monk of St. Augustine's.

[5]   'quondam cithara tacentem   'At times Apollo stirs with his lute the silent
    Suscitat musam neque semper arcum   muse and does not always draw (tendit) the bow.'
    Tendit Apollo.'—HORACE, Bk. II. Ode X.

Crossbowmen were much encouraged by Richard I. (1189–1199) in his army, and were employed by this king in his crusade in Palestine, and in his wars with France. Richard I. probably in great measure re-introduced the crossbow and caused it to become a common arm in warfare, as for a number of years previous to his reign it had suffered in popularity owing to the papal decrees against its use. Brompton[1] wrote of Richard I.: 'Truly this kind of shooting, already laid aside, which is called crossbow shooting, was revived by him, when he became so skilful in its management that he killed many people with his own hand.'

Both 'Le Breton'[2] and 'Guiart'[3] attribute the re-introduction of the crossbow, as a popular weapon in warfare, to Richard I.

Commenting on the death of Richard, Le Breton writes: 'Thus perished by the crossbow, which the English account dishonourable, King Richard, who first introduced the crossbow into France.'

After the death of Richard I., King John and Henry III. employed considerable numbers of mercenary crossbowmen in their armies, both mounted and on foot. At the second battle of Lincoln, 1217, in the civil war of 1215–1217, the relieving force sent to Lincoln consisted of 400 knights, a number of foot-sergeants, and 317 crossbowmen.

At the battle of Taillebourg, 1242, when Henry III. was defeated by Louis IX., the former had in his army 700 crossbowmen. Shortly after the death of Henry III., in 1272, the longbow came to the front in England, and for this reason the crossbow gradually became less popular with English soldiers and commanders, and continued to decline in favour as the powers of the longbow were realised.

When Henry V. led his army of some 30,000 men from England, in August 1415, he had rather less than 100 crossbowmen among his forces. At the battle of Agincourt, October 25, 1415, he had only about 8,000 troops at his disposal, owing to disease, and also to having left a garrison at Harfleur. Among these 8,000 men, there were but 38 crossbowmen, as recorded in Rymer's muster-roll of the army of Henry V.[4]

Crossbows, it may be said, were in very common use in warfare on

---

[1] John Brompton, monk of Jervaulx, abbot in 1437. His chronicle records events from 588 to 1198.

[2] Guillaume le Breton, Bishop of Tours, French chronicler, born about 1170, died 1230, wrote chronicles of the history of France in the thirteenth century, and, in 1224, the prose poem *La Philippide*.

[3] Guillaume Guiart, French chronicler and soldier, wrote a history of France in verse, recording events from 1165 to 1306. He was born about 1290.

[4] T. Rymer—historian, born 1641, died 1713. He was the son of Ralph Rymer, Lord of the Manor of Brafferton, Yorkshire.

the Continent from about 1200 to 1460–70.[1] In English armies, mercenary crossbowmen were numerous till about 1300; after which period, though by no means dispensed with, especially in the defence or siege of a fortification, they were employed in smaller numbers.

In the thirteenth and fourteenth, and in the first half of the fifteenth century, crossbowmen were considered, on the Continent, to be the 'corps d'élite' of an

FIG. 18.—MOUNTED CROSSBOWMAN, WITH CRANEQUIN CROSSBOW, AND A QUARREL IN HIS HAT.

From 'Insignia Sacræ Cæsareæ Majestatis,' P. Lonicerus, 1579.

army, and were always placed in the front of the battle line. 'Balistarii semper præibant,' wrote Matthew Paris in the thirteenth century.[2]

Among English troops, crossbowmen were given a similarly honourable position till the time when longbowmen came forward at the end of the thirteenth century.

[1] The crossbow is occasionally alluded to by French chroniclers during the reign of Louis VI. (le Gros), 1108–1137. The weapon became common in France during the reign of Philip II., 1180–1223.

[2] Matthew Paris—Benedictine monk of St. Alban's, English historian, died 1259.

Near the close of the fourteenth century, the Continental crossbow had become such a costly weapon, and one of such importance in warfare, that in Spain the crossbowman was even granted the rank of a knight. The position of 'Master of the Crossbowmen,' in France, Italy and Spain, was one of great honour, and only bestowed on persons of high consequence and title. A troop of mounted crossbowmen, of special skill and courage, usually formed the bodyguard of the king, and attended him in battle. Mounted crossbowmen were largely employed on the Continent in the fourteenth, and first half of the fifteenth century, and these men were usually allowed one and sometimes even two horses apiece, besides being supplied, when on the march, with carts to carry their crossbows and quarrels.

Hand-guns commenced slowly to supersede crossbows in Continental armies between 1460 and 1470, though the latter continued more or less in favour till the close of the fifteenth century. Paolo Giovio[1] writes that 'at the entry of Charles VIII. into Rome, in 1494, there were 500 Gascons among the troops, almost all carrying crossbows with steel bows.'

Crossbowmen were employed in limited numbers (on the Continent only) till about 1515, except in France, where, according to De Montluc, (p. 39,) the crossbow was the popular weapon of the soldier, both mounted and on foot, till 1518–20.

At the battle of Marignano, near Milan, September 1515, where Francis I. defeated the Duke of Milan and the Swiss, this king had among his bodyguard 200 mounted crossbowmen who rendered signal service.[2]

In Spain, soldiers armed with crossbows were also retained till a later date than was the case in other foreign countries and states, France excepted.

The famous General Cortes had a company of Spanish crossbowmen in his army at the siege and capture of Mexico in 1521, and employed them in defence and assault as freely as he did his soldiers armed with hand-guns. The small band of about a hundred adventurous Spaniards who sailed from Panama in 1524, under Pizarro, to explore Peru, consisted of crossbowmen only. In the conquest of Peru, 1532–1533, Pizarro had, however, only about a dozen crossbowmen among his followers.[3]

Crossbowmen were finally discarded in open warfare by all Continental armies between 1522 and 1525, but were occasionally used on foreign ships of war, and in the defence or attack of a besieged town or castle, till 1530–1535.

[1] Paolo Giovio—Italian historian, born 1483, died 1552.
[2] Guillaume de Bellay.   [3] Prescott's *Conquest of Mexico and Peru.*

There can be no truth in the assertion of Francis Grose ('Military Antiquities,' 2 vols., printed 1786 and 1788) that so late as 1572, Queen Elizabeth engaged with Charles IX. of France to supply him with 6,000 mercenary troops partly armed with crossbows.

In 1572, and for nearly forty years before this date, the crossbow was practically extinct in warfare, and even the English longbow in 1572 was rapidly falling into disuse.

The large oblong prints (the original paintings were both burnt when Cowdray House was destroyed by fire in 1793) depicting (1) the siege of

FIG. 19.—SHOOTING DEER WITH THE CROSSBOW.

The figure on the left is bending his crossbow with a belt-claw,[1] whilst he holds in his mouth
the arrow he is about to use.

From MS. Gaston Phœbus. Fourteenth Century.

Boulogne by Henry VIII. in 1544, and (II) the encampment of the English forces near Portsmouth, and the engagement between the English and French fleets, July 19, 1545, show many hundred figures of soldiers carrying pikes, hand-guns, and longbows, but not one crossbowman is to be seen among the combatants represented.[2]

---

[1] See Chapter XV. for a description of the belt-claw.

[2] Probably the last occasion on which crossbows were used against regular British troops, was at the assault and capture of the Taku forts in 1860, when many of the Chinese carried crossbows. One of their repeating crossbows, with its bamboo bow, and magazine for holding arrows, is shown in fig. 169, p. 238. It was thrown away by a Chinese soldier when the allies entered the fortifications.

FIG. 21.—SHOOTING PARTRIDGES, AS THEY FEED, BY MEANS OF THE CROSSBOW AND A STALKING HORSE.

Reduced from Stradanus's 'Venationes Ferarum.'

The stalking-horse is carried by an assistant, and takes the form of a cow. A bell, such as was, and still is usually worn by cattle in Italy and other parts of the Continent, is suspended to the neck of the stalking-horse to disarm suspicion on the part of the game. The crossbows are stonebows (see note on Stradanus, fig. 12, p. 24).

The crossbow was popular on the Continent for killing deer, till about 1635, possibly in parts of Italy and Spain even later, as there are many fine examples of powerful bolt-shooting sporting crossbows of Spanish and Italian construction, which are dated between the years 1640 and 1650.

Several writers on the chase—notably Salnove—describe the crossbow as being employed for killing deer in France during the reign of Louis XIII., 1610–1643.[1]

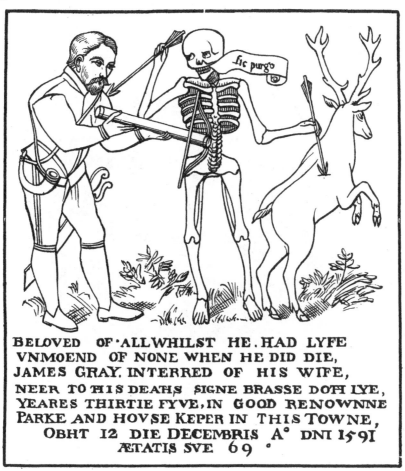

BELOVED OF·ALL WHILST HE.HAD LYFE
VNMOEND OF NONE WHEN HE DID DIE,
JAMES GRAY. INTERRED OF HIS WIFE,
NEER TO HIS DEAHS SIGNE BRASSE DOTH LYE,
YEARES THIRTIE FYVE, IN GOOD RENOWNNE
PARKE AND HOVSE KEPER IN THIS TOWNE,
OBHT 12 DIE DECEMBRIS A° DNI 1591
ÆTATIS SVE 69 ·

FIG. 20.—TABLET IN HUNSDON CHURCH.

This illustration is reduced from a brass tablet in the church at Hunsdon, in Hertfordshire, and is of an allegorical nature. It was erected to the memory of James Gray, a keeper of Hunsdon Deer Park, who died in 1591, and shows us that the crossbow was used for killing deer in England at the end of the sixteenth century.

From ' English Deer Parks,' by Evelyn Shirley. 1867.

The same weapon held, to some extent, a similar position in England till, at all events, 1621, for there is a full report extant of the commission formed of twelve bishops who, at the request of James I., inquired into the death of

[1] Salnove, Robert de, *La Venerie Royale*, en iv. parties. Paris, 1645.

Peter Hawkins, a park keeper at Bramshill in Hampshire, who was accidentally slain in 1621 by Archbishop Abbot of Canterbury, with a bolt from a crossbow which the prelate had aimed at a stag.

The archbishop, being of sedentary habits, had been ordered by his doctors to take exercise to improve his health, and at the time of the accident he was, for this purpose, enjoying the hospitality of his friend Edward, eleventh Lord Zouche. Bramshill, with its, to this day, splendid mansion and wild park, was at that time the residence of Lord Zouche, though it has since passed into the possession of the Cope family, its present owner being Sir Anthony

FIG. 22.
CROSSBOWMAN WITH A STONEBOW.
From Stradanus.

Cope, thirteenth baronet. The legend of the 'Mistletoe Bough' is said to have originated at Bramshill.

When King James heard of the archbishop's misfortune, he remarked, 'An angel might have miscarried in this way.'

Abbot's Hospital at Guildford in Surrey, is a memorial of the Archbishop's benevolence.

Early in the sixteenth century, the double-stringed stonebow was introduced, and at once became very popular with sportsmen, ladies, foresters and keepers, as a means of obtaining game-birds, pigeons, hares and rabbits.[1] Towards the middle of the seventeenth century, bolt-shooting crossbows had, however, chiefly become articles of amusement and were much used at the target, though they were employed for killing small animals till about 1720.

About 1760 the stonebow, which had always been more or less in favour, was improved in strength and accuracy, and between 1810 and 1820 it was brought to great perfection, and has since been known as the bullet crossbow.   Chapter XXXVII.

Competitions at the target with a small bolt-shooting crossbow, have for several centuries been a common recreation in parts of the Continent, especially in North Germany and Belgium.   The crossbow now used at the target in Belgium, an excellent weapon of its kind, is described in Chapter XLII.

It is a curious fact that the figure of a bird made of wood, and called the 'Popinjay,' is still set up as a mark for the modern crossbow shooters of the Continent, the name Popinjay being applied to the same form of target so long ago as the early years of the fourteenth century, fig. 161, p. 225.

---

[1] Shakespeare alludes to the stonebow.   In *Twelfth Night, or What you Will*, act ii. scene 5, Shakespeare makes Sir Toby exclaim 'O ! for a stonebow ! to hit him in the eye.'

PART II

THE CONSTRUCTION AND MANAGEMENT

OF

CROSSBOWS

MEDIÆVAL

FIG. 23.—BALISTARIUS.

A crossbow maker in his shop with the stock of a crossbow in his hand.
Engraved by Jost. Amman.

From a Work on ' Mechanical Arts,' by Hartman Schopper, 1568

CHAPTER X

*THE PRIMITIVE CROSSBOW, WITH A BOW OF SOLID WOOD,
WHICH WAS BENT BY MANUAL POWER ONLY*

THE earliest crossbow doubtless had its bow formed of one stout piece of tough wood, such as ash or yew. It was bent by drawing its string to the catch of the lock by means of the hands alone.

The feet were pressed against the centre of the bow to gain a leverage, one foot on each side of the stock. As the primitive crossbow had no stirrup, the back of its bow could be placed close to the ground, for the purpose of placing the feet upon it preparatory to drawing its bow-string.

Fig. 24, next page, shows a crossbowman bending his weapon in this manner.

These simply constructed crossbows may be recognised in illuminated missals by the absence of a stirrup, and by the length, thickness and roughness of their bows (as if wrapped outside with cord to strengthen them). This thickness, their size and rough outline, and especially the absence of the stirrup, plainly show that their bows could not have been of steel, or even of composite construction.

It will here be interesting to give the description of the crossbow of about the time of the first Crusade, as written by Anna Comnena, who attributes its invention to the French.[1] This authoress not only gives us an accurate account of the weapon, but also tells us when it was first seen (in reality re-introduced) in warfare. She writes : ' It is a bow of a kind unknown to the Greeks and to the Barbarians. This terrible weapon is not worked by drawing its cord

---

[1] Princess Anna Comnena, b. 1083, d. 1148, daughter of Emperor Alexis I., wrote the *Alexiad* (the history of her father, in fifteen books). As Anna Comnena was only sixteen years of age in 1099, she could not, prodigy though we know she was, have been the authoress of the *Alexiad* if it was finished in 1099, as stated in works of reference.

In 1118, Anna was banished from court by her brother for intriguing against him. The history of her father, she tells us in her preface, was compiled to console and occupy her during her banishment. The *Alexiad* must, therefore, have been produced between 1118 and 1148.

The fact that Anna refers to the crossbow as a novelty, shows us, from our knowledge of its antiquity, that its common use in warfare had been discontinued for many years previous to the first Crusade. There is, however, sufficient evidence to prove that crossbows were carried by the Normans at the invasion of England in 1066, p. 45.

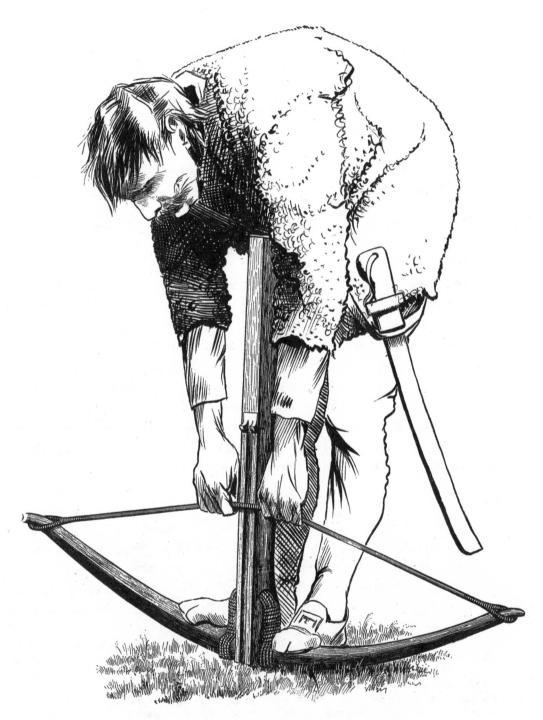

FIG. 24.—PRIMITIVE CROSSBOW WITHOUT A STIRRUP.

FIG. 25.—PRIMITIVE CROSSBOW WITH A STIRRUP.

with the right hand, and holding it with the left hand. The user rests both his feet against the bow, whilst he strains at the cord with the full force of his arms. It has a semicircular groove which reaches down the middle of the stock. The missiles, which are of various kinds, are placed in the groove, and propelled along it by the released cord. When the cord is released, the arrow leaves the groove with a force against which nothing is proof. It not only penetrates a buckler, but also pierces the man and his armour through and through.'

In course of time, the metal stirrup was fitted to the fore-end of the stock of the crossbow, as a more convenient and powerful method of bending the bow than the original one of resting the feet against the bow itself, as in fig. 24. The stirrup was the same shape as, and was no doubt suggested by the stirrup of a saddle. The crossbowman placed one foot (in the case of the larger weapons both feet) in the stirrup of his crossbow, and in this way held its stock tight to the ground, in order to resist the pull of his hands on the bow-string, fig. 25, previous page.

In military records of the thirteenth and fourteenth centuries, I find many allusions to bolts for crossbows of ' one foot ' and bolts for those of ' two feet.'[1] From this it would appear that the bolts, or the weapons for which the bolts were required, were respectively one foot and two feet long. The explanation is, that the words ' one foot ' and ' two feet ' refer to the power of the crossbows, the lengths of which were, of course, much more than one or two feet.

The larger crossbow of the period, known as ' Arbalista ad duos Pedes,' could only be strung by the soldier inserting both his feet in its stirrup, the stirrup being made wide enough for him to do this, so that he might utilise his entire weight to resist the strain exerted by his arms when bending his bow.

The smaller crossbow, known as ' Arbalista ad unum Pedem,' was lighter and of less power. For this reason, sufficient resistance was obtained by the man who used it placing one foot in its stirrup when he stretched its bow-string, the stirrup being duly shaped to this end.

Bolts for crossbows of ' two feet ' referred, therefore, to the heavier missiles that were shot from the larger weapon, and bolts for crossbows of ' one foot ' referred to the lighter shafts intended for the less powerful crossbow.

When a crossbowman bent his bow with his hands alone, as in the case of the weaker weapons, he wore a leathern guard on each hand to

---

[1] October 20, 1301. The king wishing to strengthen the town of Linlithgow, commands the Treasurer and Barons to send there six crossbows à tour with 2,000 quarrels, also twelve crossbows of two feet and 3,000 quarrels, and 5,000 quarrels for crossbows of one foot.—From *Calendar of Documents relating to Scotland*, No. 1250, Edward I. In 1328, Edward III. orders the Sheriffs of London to supply for the defence of the Channel Islands ' a hundred arcubalisti ad Pedem, and twenty arcubalisti ad Troll' (Rymer's *Fœdera*, iv. 367).

protect his fingers from being cut. These small leathern guards just covered the insides of the fingers when the latter were hooked over the bow-string. The pieces of leather were retained in position, when in use, by placing the thumbs through holes in their ends.

The primitive crossbows which were strung in this manner could have been of little power in comparison with those later ones which required mechanical aid to draw their strings, such as crossbows with composite, or steel, bows. The former may, however, have been effective at a time when the bow was little used in Continental warfare, and before the powerful English longbow came to the front.

The primitive crossbow was, probably, not only a more accurate arm than the ordinary bow of its period, but also one of a more dangerous nature, as it projected a much heavier arrow than that of a bow.

The fact that the primitive crossbow (see Anna Comnena, p. 57) required the utmost strength of both arms to pull back its string, proves that it must have discharged its missile with considerable force, a force, perhaps, sufficient to penetrate, at a short range, leathern jackets or even coats of mail.

CHAPTER XI

THE THIRTEENTH AND FOURTEENTH CENTURY CROSSBOW, WITH A COMPOSITE BOW (OF YEW, HORN AND TENDON), WHICH WAS BENT BY HAND, OR BY A THONG AND PULLEY, OR BY A METAL CLAW ATTACHED TO THE CROSSBOWMAN'S BELT

ARBALESTE DE COR ET D'IF

WHEN the bow of a crossbow was shaped out of a single piece of wood, as in the earliest weapons of the kind (figs. 24, 25), it must always have been liable to break or warp, or take a 'set,' after being for some time in use. For this reason, the crossbow with a beautifully constructed composite bow, composed of horn or whalebone, yew and tendon, superseded the weapon with a solid wooden bow.

The crossbow with a composite bow is said to have been brought to Europe from the East by the Saracens, during the Crusades of the twelfth century, and through them popularised on the Continent. At the time of the Crusades, and for many years after, the Saracens were famed for their construction of crossbows. In a list of crossbow makers compiled by Baron de Cosson, the name of 'Peter the Saracen' is the earliest he can find mentioned, this man being maker of crossbows to King John of England in 1205.[1]

It is likely that the weapon used by the Normans in the conquest of England, had a stout bow of solid wood. In the time of Richard I., however, 1189–1199, this king probably hired crossbowmen with composite bows formed of horn, wood and tendon; crossbows with steel bows being of later date.

In support of the latter contention, I may quote Justiniani,[2] who writes that in 1246 (or 47 years later than Richard I.) '500 Genoese crossbowmen whose crossbows had bows of horn,[3] were sent against the Milanese, and that

---

[1] *Close Rolls of King John*. Bentley. 'Excerpta Historica 395.'
[2] Bernardo Justiniani—Italian historian, born 1408, died 1489.
[3] Composite ones of horn, wood and tendon.

each Genoese who was captured by the enemy was deprived of an eye and an arm, in revenge for the loss of life inflicted by his crossbow.'

The composite bow, as applied to the crossbow, was of rather clumsy appearance, and, unless closely examined, might easily be mistaken for a bow of wood in one piece. The composite bow was, however, light, elastic and fairly powerful, far more so than a bow of solid wood, and before the

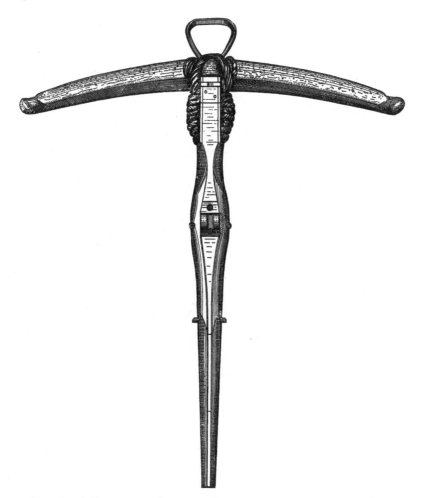

FIG. 26.—A FIFTEENTH CENTURY CROSSBOW WITH A COMPOSITE BOW
WHICH WAS BENT BY A CRANEQUIN. (German.)

days of longbows and steel crossbows, it was probably an effective weapon in warfare.

These composite bows may be recognised in illustrated manuscripts by their short length,[1] great thickness, and smooth outlines, by the presence of a stirrup on the fore-end of the stock in the earlier weapons, and especially by

---

[1] The composite bow of a crossbow was sometimes as much as $2\frac{1}{2}$ in. wide and $1\frac{1}{2}$ in. thick, though in length seldom over 2 ft. 5 in., more often 2 ft. 3 in. or 2 ft. 4 in.

the fact that their bows have no curve like those of solid wood or of steel, but are nearly straight before they are bent, fig. 26, previous page.

They were formed of horn or whalebone, yew, tendon and glue. The heart, or core of the bow was composed of about twenty thin flat strips of horn or whalebone, placed side to side and glued one to the other into a solid block, the thin edges of the pieces being respectively towards the back of the bow and its string. That is to say, the twenty thin pieces bent collectively edgewise, and not flatwise when the bow was used. At the back and front of the longitudinal block of horn or whalebone, which formed the mainspring or heart of the bow, a strip of yew was attached. A thick coating of the tendon of an animal[1] was then moulded all round the horn and yew, in order to hold these parts together, and to add to the power of the bow by its great elasticity. The bow was finally thickly coated with glue, or skin covered with varnish. This was done as a means of resisting damp from the outside and to keep the inside parts of the bow soft and pliable, by hermetically sealing them from contact with the atmosphere.

I need scarcely point out that a bow of solid horn would be useless in a crossbow. Such a bow, being comparatively very short in relation to its length and substance, would be sure to fracture. The word 'horn' merely referred to the heart or backbone of the composite bow, to distinguish it from a solid wooden bow, or from a steel one.

Victor Gay, for instance, in the 'Glossaire Archéologique,' gives an extract from Gilles le Bouvier dated 1455.[2] Le Bouvier writes as follows :— 'These people (Bavarians) are good crossbowmen on horseback and on foot, and shoot with crossbows of horn and sinew, which are good and strong and do not break. Those of horn do not break when they are frozen, for the colder it is the stronger they are.'

The smaller and more ancient crossbow with a composite bow was probably strung by the hands alone, as described by Anna Comnena, p. 57. The larger, such as those of the thirteenth and fourteenth centuries, by means of a leathern thong and a pulley, or by the aid of a claw attached to the crossbowman's belt, and in the case of mounted men by a goat's-foot lever.

---

[1] This ligament or tendon, was the 'ligamentum colli,' or pack-wax, of the ox or horse, and differs from the other ligaments, in that it possesses great elasticity. The 'ligamentum colli,' or ligament of the neck, supports the heavy head of the horse in an erect position, without the least muscular effort on the part of the animal. If, however, the horse lowers its head to the ground to feed or drink, this ligament is so elastic that it then lengthens fully two inches. When the horse commences to elevate its head again, after it has ceased to feed, the ligamentum colli at the same time contracts, and thus enables the animal to lift its head without any exertion. The mediæval crossbowman cleverly utilised this very powerful and elastic ligament as a means of adding strength to his bow. It was also sometimes used for the skein of the Roman catapult.

[2] Gilles le Bouvier—French chronicler, born 1386, died about 1457.

In manuscripts of the thirteenth, fourteenth and first half of the fifteenth century, I find many allusions to soldiers armed with horn crossbows.

Doubtless on the coasts of Scandinavia and North Germany, the chief home of these composite crossbows after the time of the Crusades, whalebone was easily obtainable, whilst in other parts of the Continent, the pieces which formed the heart of the bow, were made from the straightened horn of an animal.

This ancient form of crossbow with a composite bow, survived in an improved form in Scandinavia and in the north of Europe, as a weapon of sport and war, till about 1460, or for nearly a hundred years after the far superior crossbow with a thick steel bow and a windlass had been in use in France, Spain and Italy. Some of these later weapons were made so strong in the fifteenth century, that after the invention of the powerful cranequin for bending steel bows, this apparatus was also employed for bending the composite bow.

Several of the larger crossbows with composite bows, to be seen in the museums of North Germany, have the steel cross-pin projecting on each side of the stock, some six or seven inches behind the catch for the bow-string, which shows beyond question that a 'cranequin' was applied to bend their bows, fig. 26, p. 63.

CHAPTER XII

HOW THE BOW OF THE PRIMITIVE CROSSBOW WAS ATTACHED TO THE STOCK BY A BRIDLE OF CORD OR SINEW

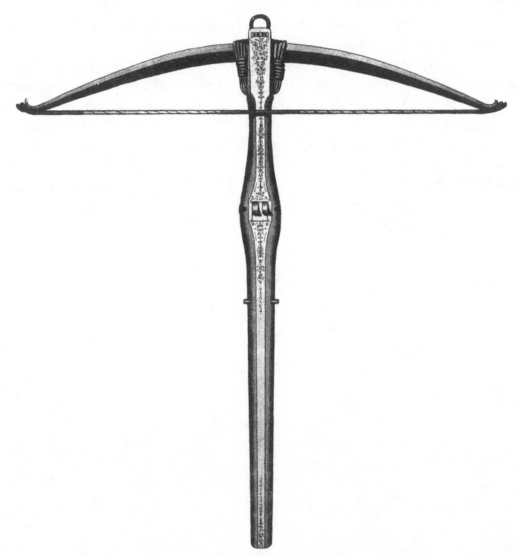

FIG. 27.—A SIXTEENTH CENTURY SPORTING CROSSBOW THAT WAS BENT WITH A CRANEQUIN, AND WHICH HAD A STEEL BOW ATTACHED TO THE STOCK BY A BRIDLE OF SINEW. (Spanish.)

IN the primitive crossbow—in which class the earlier weapon with a composite bow may be included—the bow was attached to the stock by means of an ingenious bridle made of cord or sinew.   This bridle proved a light and very strong method of securing a wooden, or a composite bow to its stock.   It not only greatly lessened the jar caused to the stock, by the rebound of the bow when the crossbow was discharged, but also held the bow in its grasp without causing the damage to it that would arise from metal clamps.

Though this bridle of cord or sinew is seldom seen in the large military crossbow with a heavy steel bow, it was commonly used in the smaller weapons with steel bows which were employed for sporting purposes in the sixteenth century, fig. 27.

THE BRIDLE OF SINEW WHICH WAS OFTEN USED FOR ATTACHING THE BOW OF A CROSSBOW TO ITS STOCK (FIG. 28, NEXT PAGE).

I. Fig. 28.   The saddle, or piece of hard wood—along its flat side of the same breadth and curve as the bow—which rested upon the centre of the back of the bow.   When the bow was in position, the hollows in this piece projected just clear of either side of the stock, and held from slipping the wrapping which secured the bow and formed the bridle.

II. Fig. 28.   The bow fixed to the stock.   Front and side view.

A, is one end of the saddle.   B, is the bow.   C, is the wrapping or bridle. D, is the oval hole in the stock through which the wrapping forming the bridle is threaded.   (The hole for the wrapping, and the opening for the bow and its saddle, were concealed by little bunches of coloured wool.)

III. Fig. 28.   The wrapping as first put on, and before it is bound together at E E, on each side of the stock, in order to tighten it and thus fix the bow. The wrapping, usually consisting of deer or other sinew softened in water, was firmly wound over the projecting hollows of the saddle A, which rested upon the back of the bow.   It was passed ten or twelve times, to and fro, through the oval hole D in the stock, and alternately over each end of the saddle. The separated halves of the wrapping (E E, III. fig. 28) were then forcibly drawn together on each side of the stock by another length of strong pliable sinew, as seen in II., fig. 28.

The wrapping, of course, gradually tightened throughout, as its side strands were pulled up close, with the result that the bow was forced immovably up to the stock.

––––––––––

When the bridle of sinew was dry and set, it became almost as tight and rigid as an iron screw clamp.

I have had crossbows with steel bows that were secured in this way over

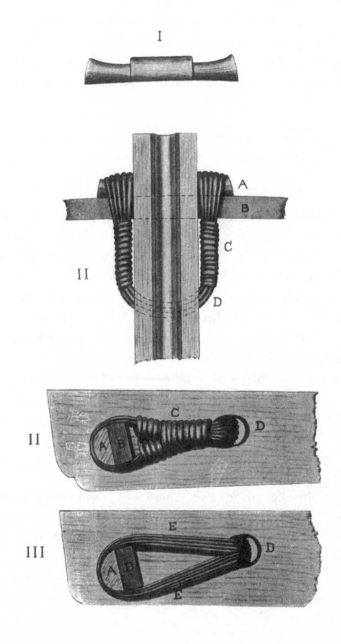

FIG. 28.—THE BRIDLE OF SINEW WHICH WAS OFTEN USED FOR SECURING THE
BOW OF A CROSSBOW TO ITS STOCK.

three hundred years ago, and which I could not knock out of their stocks with a heavy hammer, without first cutting through the bridle of sinew that still held them fast in their original positions.

The lock of the primitive crossbow—with its ivory tumbler (known as the nut) and long trigger was precisely the same as the lock described in Chapters XX, XXI. This simple form of lock for holding and freeing the bow-string, was common to all bolt-shooting crossbows till about the middle of the sixteenth century.

CHAPTER XIII

HOW THE CROSSBOWMAN PLACED THE BOLT OF HIS CROSSBOW ON THE STOCK OF HIS WEAPON, SO AS TO GIVE THE BOLT A FREE LOOSE AND CORRECT FLIGHT

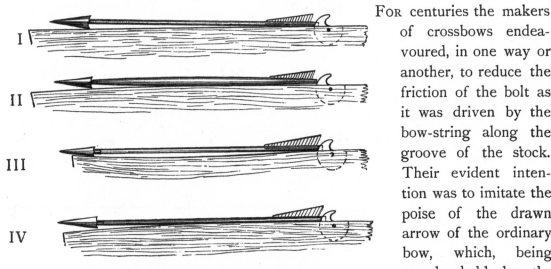

FIG. 29.—BOLTS FOR CROSSBOWS, AND HOW THEY WERE ARRANGED ON THE STOCK.

FOR centuries the makers of crossbows endeavoured, in one way or another, to reduce the friction of the bolt as it was driven by the bow-string along the groove of the stock. Their evident intention was to imitate the poise of the drawn arrow of the ordinary bow, which, being merely held by the fingers at one end, and lightly balanced on the archer's hand at the other, was but slightly retarded by friction when released.

These efforts of the crossbow-makers to decrease the friction of the bolt of the crossbow against the stock, as the bolt was driven forward by the bow-string, may be traced in all mediæval crossbows or their bolts. The methods employed are described in fig. 29. See also fig. 88, p. 137.

I. Is the side view of the fore-end of a crossbow stock, which in this case is straight. The head of the bolt, however, lifts up the shaft, and in this way prevents its frictional contact with the stock.

II. The stock, also intended to be straight. As the shaft of the bolt is enlarged near its centre of length, it rests at that part only on the stock, with the result that friction is reduced when the projectile is shot forward.

III. The stock with its middle part slightly hollowed out, so that the bolt—as may be seen—only rests at its point and at its butt on the stock.

IV. Here the stock is sloped downwards, from a point near the balancing point of the bolt. The bolt leaves the stock, therefore, without much friction against it.

This last system gave a free and easy quittance to the bolt, and is one that is even now applied to some of the modern target crossbows of the Continent, fig. 145, p. 207.

Chapters XIV–XVII describe the methods employed in the twelfth, thirteenth and fourteenth centuries, for bending crossbows with light steel bows or composite ones, which, though of no great strength, were too powerful to be bent by manual power.[1] These methods were,

I. The cord and pulley.   III. The screw and handle.
II. The claw and belt.   IV. The goat's-foot lever.

As described in Chapter X, the primitive crossbow, with its bow of one piece of solid wood, was bent by hand without other aid.

It was not long, however, before a stronger bow was fitted to the crossbow, or one that could not be bent without some form of lever, and which was, of course, much more effective than the weaker kind of bow previously used.

It is uncertain when crossbows were commonly made with steel bows, (instead of with wooden ones, or with composite ones of wood and horn,) probably not before the middle of the fourteenth century. See notes on crossbows at Crécy, p. 5.

The windlass, suggested no doubt by its application to the siege engine that cast javelins, was not applied to the crossbow till the latter half of the fourteenth century. Being of great power, the windlass allowed of a far stronger steel bow than was possible previously to the introduction of this kind of winder for drawing the string of a crossbow.

The levers designed for bending the crossbows in use before the perfecting of the windlass, were of no great force ; hence the bows to which these devices were adapted, were only of moderate strength.

As it was an evident advantage to the crossbowman to carry a weapon with as powerful a bow as possible, it was imperative that he should contrive some mechanism for pulling back his bow-string, when his bow had developed into one that was too unyielding for him to bend by hand alone.

---

[1] Though these appliances are not mentioned before the thirteenth century, it is probable that such simple devices as the cord and pulley, or the claw and belt, were used for bending a crossbow shortly after the invention of the weapon.

The early appliances invented by the crossbowman for this purpose I will now describe.

Their introduction may be taken to date from the period when the crossbow bent by hand was superseded, and they were, one or the other, commonly employed till the time when the heavy crossbow appeared which was bent by a windlass.

I should add, however, that some of these primitive methods, notably the claw and belt, were retained till the close of the fourteenth century for bending the lighter crossbows carried by hunters, whilst the goat's-foot lever was used in both military and sporting weapons of medium strength till a later date.

CHAPTER XIV

THE VARIOUS CONTRIVANCES EMPLOYED IN THE THIRTEENTH AND FOURTEENTH CENTURIES FOR BENDING THE BOWS OF CROSSBOWS WHICH WERE TOO STRONG TO BE BENT BY MANUAL POWER ONLY

THE CORD AND PULLEY

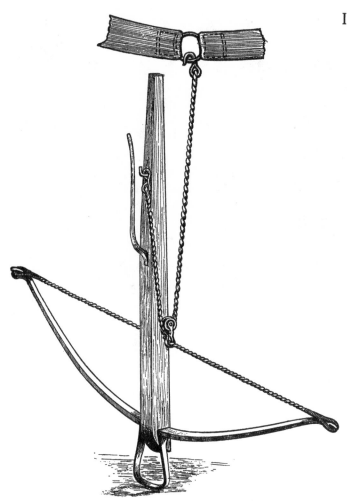

FIG. 30.—CROSSBOW WITH CORD AND PULLEY.

IN fig. 31, next page, we see two figures of crossbowmen taken from the great picture by Antonio Pollajuolo, of the martyrdom of St. Sebastian, 288. This picture was painted in 1475, and is now in the National Gallery.

The crossbows depicted in fig. 31 have composite bows of wood and horn, and their bow-strings are each being stretched by means of a cord and pulley.[1]

In this method of bending the bow of a crossbow, one end of a stout piece of cord was secured to a ring in the crossbowman's belt. The other end of the cord was passed over the wheel of a small pulley, and was then hitched to a

---

[1] Crossbows bent in this way were known as Turni Balistarii and Arbalests à tour. They are mentioned in the thirteenth century. In 1301, Edward I. sends a demand for some crossbows à tour for the defence of Linlithgow (footnote, page 60)

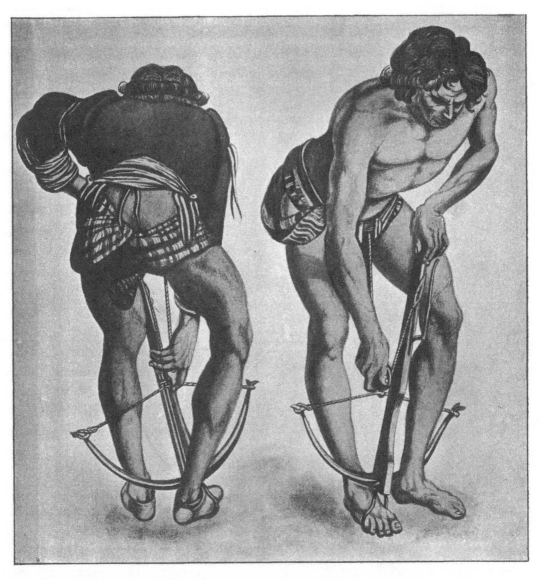

FIG. 31.—CROSSBOWMEN BENDING THEIR BOWS WITH THE CORD AND PULLEY.

Antonio Pollajuolo. 1475.

metal stud fixed below the shoulder-end of the stock of the crossbow, fig. 30, p. 73. Sometimes a thong of hide was substituted for the cord.

The single claw of the pulley was hooked over the bow-string close to its centre, and guided in its course along the stock, and also held from slipping sideways, by the fingers of the crossbowman, fig. 31.

As the crossbowman straightened his legs and body from their stooping posture, he naturally applied considerable force to his cord and pulley. He was thus able, quickly and easily, to draw the string of his crossbow towards him along the groove of its stock, till it finally caught on the nut that held it fast till the trigger was pressed. The cord and pulley being then removed from the crossbow, the weapon was ready for use.

The crossbowman placed his foot through the stirrup of his crossbow, in order to hold its stock firmly to the ground, so that it might resist the strain he applied to the bow-string.

This was a rapid and effective manner of bending the bow of a crossbow which, though too strong to be bent by hand alone, was not powerful enough to require a windlass for the purpose.

The system of a cord and pulley was probably the most ancient of all devices for bending crossbows, and is one that is rarely shown in illustrated manuscripts of a later date than the end of the first quarter of the fourteenth century.

FIG. 32.—CORD AND PULLEY.

FIG. 33.—CROSSBOWMAN WITH A CLAW
FOR BENDING HIS CROSSBOW AT-
TACHED TO HIS BELT.

FIG. 34.—CROSSBOWMAN BENDING HIS
CROSSBOW WITH A BELT-CLAW.

CHAPTER XV

THE VARIOUS CONTRIVANCES EMPLOYED FOR BENDING THE BOWS OF CROSSBOWS (Continued)

THE CLAW AND BELT

A METAL claw, either single or double pronged, swinging to a waistbelt, seems to have been a popular method in the fourteenth century of bending the bows of military and sporting crossbows of the weaker kind.

Figs. 33 and 34 are taken from Viollet-le-Duc.[1] They plainly show how the claw was applied to bend the crossbow. The claw was attached to the cross-bowman's leather belt, and was either suspended therefrom by a ring fixed to its shank, or the upper end of its shank was crooked, so as to hook over the belt. As the leverage to bend the crossbow was in this case exerted by

FIG. 35.—BELT AND CLAW.

the direct pressure of the leg, it was of a much more powerful nature than the arrangement of cord and pulley described in Chapter XIV.

When the crossbowman wished to bend his crossbow, he held its stock in an upright position, with its grooved surface next him and its stirrup directed downwards, fig. 34. He hooked his metal claw over the centre of the bow-string, the stirrup at the fore-end of the crossbow being then about a foot above the ground, fig. 34. The crossbowman now raised his right foot and placed it in the stirrup. He then straightened his bended leg, and in this way forced

---

[1] Viollet-le-Duc, *Dictionnaire raisonné du Mobilier français*, Paris, 1855–1875.

his crossbow downwards, fig. 34. The bow-string, being meanwhile hooked to the claw which was fastened to the belt, was restrained from following the movement of the crossbow, as the latter was pressed toward the ground. The bow-string was, therefore, forcibly drawn along the stock of the crossbow, till at length it slipped over the catch of the lock.

In Gaston Phœbus we find some excellent representations of hunters bending their sporting crossbows with the belt claw. All the crossbows shown in these pictures appear to have thick bows of composite make, *i.e.* of wood and horn. Bows of steel would be of far lighter construction than those

FIG. 36.—SHOOTING A WILD-BOAR WITH CROSSBOWS.
One of the hunters is bending his crossbow by means of a belt-claw and his foot.
From MS. Gaston Phœbus.   Fourteenth Century.

sketched in Gaston Phœbus. Not only do we see in these illuminations the crossbowman bending his bow with the claw, but he is also shown in the act of shooting, with the claw hanging from his belt and ready for use, figs. 36, 37.[1]

NOTE ON GASTON PHŒBUS

Gaston III., Count de Foix and Vicomte de Béarn, surnamed Phœbus, was born in 1329 and died in 1391. He was a brave and celebrated knight of remarkable personal beauty, of great wealth and position, and above all things devoted to hunting. He married the daughter of Philip VI., King of France. Gaston wrote a work on the chase in two parts; the first or theoretical part exists

---

[1] See also fig. 19, p. 49.

only in manuscript, the second and more practical part was printed at Paris about 1507.

This work on the chase may justly be considered the most famous ever written on the subject, and is one from which mediæval authors, for some two hundred years, purloined their information on hawking and hunting.

Gaston commenced his book in 1387, and completed it in four years, just before his death from apoplexy after returning from a day's hunting.

Nineteen manuscript copies of the work are known to exist; thirteen of these are in the British Museum Library, and three in the Bodleian.

FIG. 37.—SHOOTING IBEX WITH THE CROSSBOW.
The hunter may be here seen with the claw for bending his crossbow attached to his belt.
From MS. Gaston Phœbus. Fourteenth Century.

The later reproductions are, however, illuminated in accordance with the ideas of their transcribers.

The more ancient copies of Gaston are, therefore, the most interesting in regard to their illustrations, though these are very roughly drawn.

Count Gaston was a patron of Sir John Froissart. Froissart ('Chronicles,' vol. iii.) gives a long and graphic account of his visit to Gaston at his castle of Orthès in France, and describes the luxury and splendour of his court, the immense retinue of servants whom the Count had to attend him at home and in the field, and the many hundred hounds he kept for use in the chase.

At the date of his visit, 1388, Froissart writes :

'Count Gaston was at this time 59 years old, and I must say that, although I have seen very many knights, kings, princes and others, I have never seen any one so handsome . . . he was so perfectly formed one could not praise him too much. . . . There were knights and squires to be seen in every chamber, hall and court, conversing on arms and amours. Everything honourable was there to be found, all intelligence from distant countries was there to be learnt, for the gallantry of the Count had brought visitors from all parts of the world.'

This eulogy of the courtier Froissart is rather discounted by the fact that Gaston was a tyrant, a voluptuary and the murderer of his own son !

FIG. 38.—A CROSSBOWMAN BENDING HIS BOW WITH A BELT-CLAW, AND THEN
AIMING HIS CROSSBOW.

From Viollet-le-Duc.

CHAPTER XVI

THE VARIOUS CONTRIVANCES EMPLOYED FOR BENDING THE BOWS OF CROSSBOWS (*Continued*)

THE SCREW AND HANDLE

As the crossbowman gradually increased the strength of his bow, with a view to acquiring a longer range and the use of a heavier missile, he, in course of time, required for drawing his bow-string some contrivance which was of more power than the cord and pulley or the belt claw.

For this purpose, he devised a rough form of screw-jack, of metal, that he could attach to the stock of his weapon when he wished to bend its bow. This apparatus, though of clumsy mechanism and tedious manipulation, was far more powerful than any system of leverage previously applied to a crossbow.

Representations of the screw and its handle are very rare in illuminated manuscripts of mediæval times, though crossbows that were bent in this manner, and which had steel bows, are frequently mentioned in the latter half of the fourteenth century.

This screw and handle bender for crossbows, must not be confused with the rack introduced in the fifteenth century, known as a 'Cranequin.' It is probable, however, that the screw here described suggested the much more convenient 'Cranequin' that succeeded it.

Fig. 39 is from a Froissart manuscript of the early part of the fifteenth century, and shows a

FIG. 39.—CROSSBOWMEN.

From Manuscript, ' Froissart's Chronicles.'

crossbowman bending his bow with a screw. He is kneeling on the ground, and is engaged in turning the handle that draws back the metal rod which stretches his bow-string.

Fig. 40 is from Valturius, and represents a crossbow with its handle and screw.[1]

FIG. 41, OPPOSITE PAGE, SHOWS HOW THE SCREW WAS APPLIED
TO BEND A CROSSBOW

The long shank of the screw rod was of a smaller diameter than, and was able to pass easily to or fro through, the longitudinal hole in the raised part of the stock of the crossbow.

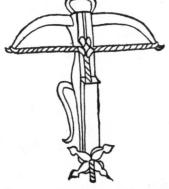

The screw inside the centre hole of the handle, fitted the screw part of the rod.

When the crossbowman wished to bend his bow, he pushed the rod through the long hole in the stock of his crossbow, and at the same time hooked the claw of the rod over his bow-string. He then screwed the handle on to the end of the rod, where it projected slightly beyond the extremity of the stock, fig. 41.

FIG. 40.—SCREW AND HANDLE
CROSSBOW.

From Valturius, Edition 1472.

He now turned the handle. The pressure of the revolving handle against the end of the stock, caused the rod to pull back with its claw the bow-string, till the latter was at length secured by the catch of the lock. The bow being bent, the handle was quickly reversed and in this way spun off the rod.

The rod was then loose, and was pushed forward to remove it from the stock and bow-string, and the crossbow was ready for use.

The end of the stock of the crossbow had a steel cap, to protect this part from being worn away by the friction of the handle, as well as to assist the handle to work smoothly when it was being turned round by the cross-bowman to bend his bow.

[1] Valturius Robertus, living at the end of the fifteenth century. Author of *De Re Militari*, first printed at Verona in 1472.

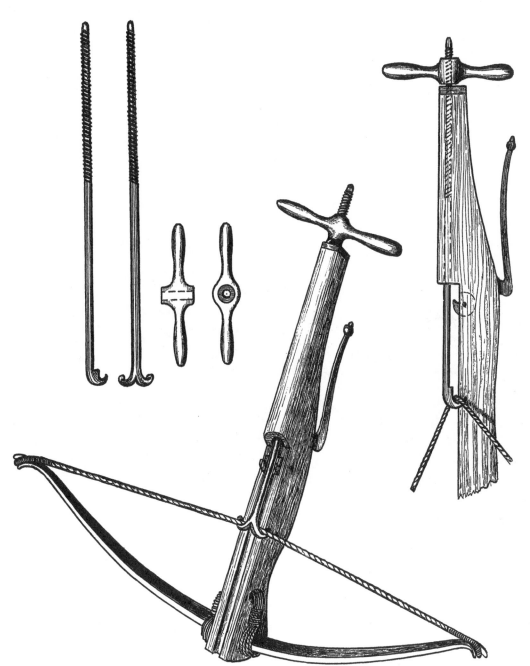

FIG. 41.—THE CROSSBOW WHICH WAS BENT BY A SCREW AND HANDLE.

CHAPTER XVII

THE VARIOUS CONTRIVANCES EMPLOYED FOR BENDING THE BOWS OF CROSSBOWS (Concluded)

THE GOAT'S-FOOT LEVER

A PIED DE BICHE——A PIED DE CHEVRE

FIG. 42.—THE GOAT'S-FOOT LEVER.

THIS apparatus for bending crossbows was known as a goat's-foot lever, from its supposed resemblance in outline to a hind-foot of a goat.   Though not of sufficient strength to bend a thick steel bow, or one such as required a windlass or a cranequin, the goat's-foot lever was of considerable power. Its action was easy and rapid, and could be applied on horseback.   For these reasons, the goat's-foot was carried by the mounted crossbowman in preference to any other kind of lever employed for stretching the bow-string of a crossbow of moderate power.

The crossbow which was bent by a thong and pulley, a claw to the belt, a rack and screw or a windlass and ropes, could not possibly have been used by the mounted soldier.   It is true the cranequin was employed for bending the larger crossbow carried by horsemen, but, as its mechanism was of elaborate and costly construction, it was not supplied to ordinary troops.   On the other hand, the goat's-foot lever was simple and cheap and could be made for a trivial sum by any worker in metal.

I do not find the goat's-foot lever represented till the middle of the fourteenth century, but from that time till the end of the fifteenth century it is frequently pictured and mentioned in contemporary literature,

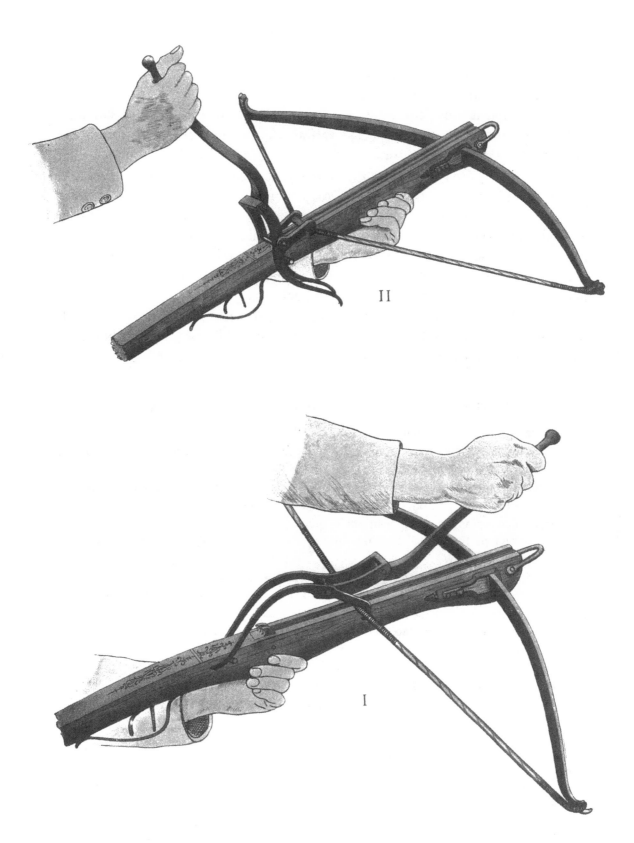

FIG. 43.—HOW THE GOAT'S-FOOT LEVER WAS APPLIED TO BEND A SMALL CROSSBOW.

and appears to have been a favourite contrivance for bending the lighter kind of military crossbow, fig. 97, p. 151.

The simplicity and convenience of this lever were so evident, that long after crossbows were discarded in warfare, it was popular for bending the steel bows of the smaller weapons used in sport or at the target—in the latter case, till as recent a date as the close of the eighteenth century.

Fig. 43, previous page, shows a seventeenth century small sporting crossbow being bent by its goat's-foot lever. From these sketches it will be realised how the mounted crossbowman held his crossbow and worked his lever. He passed his left arm through his bridle reins when in the act of bending his bow, or, in the event of his horse being well trained and steady, merely hitched the reins over the high pummel of his saddle.

Both the crossbow and its lever were fitted with small rings, by which they could be suspended to hooks fixed in the saddle of the crossbowman, when he did not require his weapon.

THE MECHANISM OF THE GOAT'S-FOOT LEVER, FIG. 44.

I. The handle, surface and side view. The handle is 10 in. long. It is $\frac{3}{4}$ in. wide at its widest part A, and tapers from a thickness of $\frac{1}{4}$ in. near its swivel end B, to $\frac{3}{16}$ in. near its small end C.

II. The fork, surface and side view. A, is the cross-pin on which the handle is hinged, and B, is the pin on which the claw-frame swings, both pins being $\frac{1}{4}$ in. in diameter. These pins are $2\frac{1}{2}$ in. from one another.

The curved parts, or prongs of the fork, are each $6\frac{3}{4}$ in. long from the cross-pin B, to their ends C–C.

The sides of the fork are $1\frac{1}{2}$ in. apart inside[1] and $\frac{3}{16}$ in. thick.

From the bend of the fork near A, to B, the sides are $\frac{3}{4}$ in. wide; they then gradually decrease in width to $\frac{3}{16}$ in. at their points C–C.

III. The claw-frame, surface and side view. This part of the lever swings loosely on the cross-pin B.

The sides of the claw-frame are $2\frac{3}{4}$ in. long and $\frac{3}{8}$ in. wide. From D to E, they are $\frac{3}{8}$ in. thick, from E to F they are $\frac{3}{16}$ in. thick. The claws are $1\frac{1}{8}$ in. apart inside The flat cross-bar G, which connects the claws, is $\frac{1}{2}$ in. wide and $\frac{1}{8}$ in. thick.

HOW TO USE THE GOAT'S-FOOT LEVER, FIG. 43, PREVIOUS PAGE.

Hook its claws over the centre of the bow-string, a claw being on each side of the stock and just clear of it.

---

[1] This width of $1\frac{1}{2}$ in. fits a stock which is $1\frac{1}{4}$ in. wide across its grooved surface. If the width of the stock of a crossbow at this part is more or less, then the width between the sides of the fork will of course vary to suit.

Place the prongs of the fork over the top of the stock, with their ends resting upon the transverse iron pin ($\frac{1}{2}$ in. thick) which projects $\frac{3}{4}$ in. on opposite sides of the stock, below the catch of the lock, I, fig. 43, p. 85.

[The ends of this pin were sometimes fitted with small revolving collars, to assist the downward slide of the fork as pressure was put on the handle of the lever.]

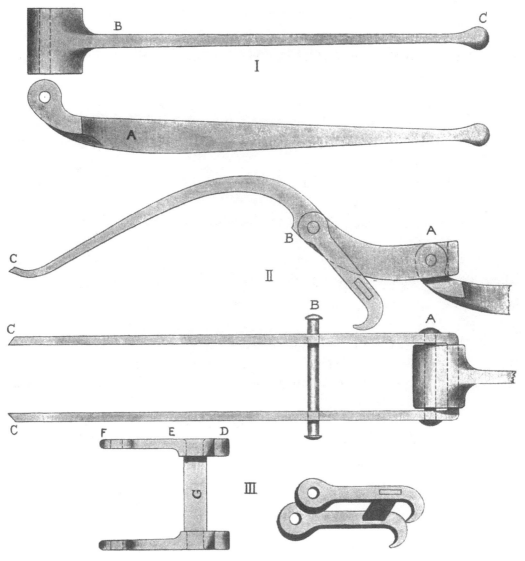

FIG. 44.—THE MECHANISM OF THE GOAT'S-FOOT LEVER.
Half full size.

Hold the crossbow in a level position with the left hand, the shoulder-end of the stock resting against the front of the right thigh. Pull the handle of the lever towards you with the right hand, II, fig. 43, p. 85, and fig. 97, p. 151.

The leverage obtained from the fork of the lever, as you pull its handle back, will enable you to stretch the bow-string to the catch of the lock smoothly and quickly.

I, fig. 43, p. 85. The lever fitted to the stock and bow-string, and ready to stretch the string over the catch of the lock.

II, fig. 43. The bow-string stretched over the catch of the lock by pulling back the handle of the lever. The lever having now no strain upon it from the bow-string, is loose, and may be removed from the stock by lifting it upwards. When not in use the handle of the lever is hinged back, so as to lie between the sides of the fork.

The goat's-foot lever I have described, was adapted to fit the small crossbow carried by mounted soldiers, as well as the light weapon employed in the chase or at the target.

In the case of foot-soldiers, however, a more powerful crossbow was used than could be managed on horseback, its goat's-foot lever being also larger, to enable it to bend the bow of the stronger weapon.

This crossbow could only be bent by resting its stock on the ground and then forcing the handle of its lever downwards with the right hand, whilst the left hand grasped a stout metal ring secured to the fore-end of the stock. Fig. 45, opposite page, shows a crossbow being bent in this way by its goat's-foot lever.

In these weapons of the foot-soldier, a lever of proportionate thickness to the strength of the bow intended to be bent, was, of course, necessary, the lever being usually about one-third longer in all its parts than the one given in fig. 44, and of suitable strength.

I should add that though these heavier crossbows were of considerable power and efficiency in warfare, they were much inferior in range and penetration to the crossbow that could not be bent by a goat's-foot lever, and which required a windlass or a cranequin for the purpose.

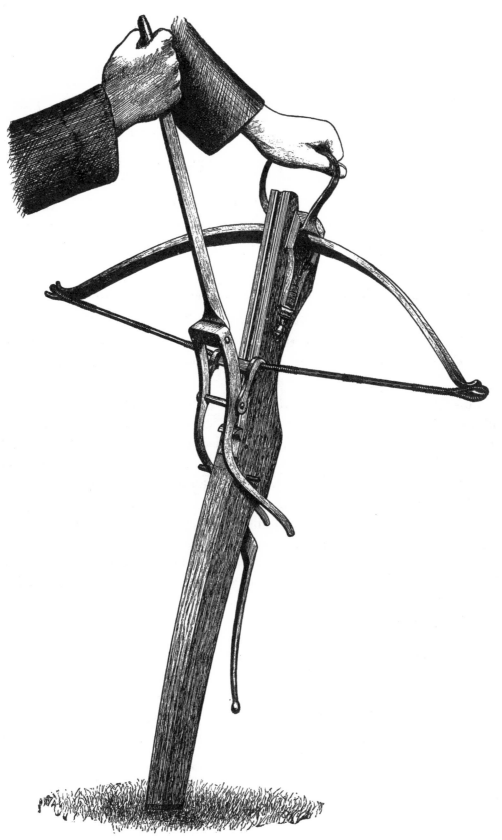

FIG. 45.—A MILITARY CROSSBOW BEING BENT BY A GOAT'S-FOOT LEVER.

CHAPTER XVIII

THE FIFTEENTH CENTURY MILITARY AND SPORTING CROSSBOW, WITH A THICK STEEL BOW, WHICH WAS BENT BY A WINDLASS AND ROPES, AND DISCHARGED A BOLT

GROSSE ARBALESTE—ARBALESTE A MOULINET—ROLLING PURCHASE CROSSBOW—
WINDLASS CROSSBOW

THERE is no evidence to show the exact period when the perfected military crossbow—which was so popular on the Continent in the fifteenth century—was first used in warfare.

This powerful crossbow, with its thick and broad steel bow and its windlass,[1] is first alluded to in contemporary accounts of battles and sieges which occurred shortly before the last quarter of the fourteenth century.[2] It is, however, probable that crossbows with steel bows were in use soon after Crécy, their bows being comparatively small and weak, and bent by the thong and pulley, claw to the waist-belt, or by goat's-foot levers.

The smaller steel crossbow was either slung upon the back of the foot-soldier, or suspended from the saddle of the mounted man.

The large military crossbow was far too ponderous to be carried by a man on horseback, nor could its bow be bent by any apparatus except its heavy windlass, a method of winding up the bow-string which would have been impossible in the case of horsemen.

---

[1] Windlass or moulinet. In one form or other, the windlass had been used for bending the bow of the Roman Balista for centuries before it was applied to the crossbow carried by hand. See Balista, Chapter LVII.

[2] In the illustrations appended to Froissart's chronicles, this crossbow is frequently shown as being used in the battles and sieges of the first half of the fourteenth century, as at Crécy for instance. The illustrations to the chronicles were drawn, however, by artists of the fifteenth century, who no doubt pictured the weapon they were then acquainted with. For instance, the illustrations showing windlass crossbows, pp. 4, 7, 20, are from fifteenth century MSS. of Froissart's chronicles. This, and the other drawings in his translation, were reproduced by Colonel Johnes, 1803-5, chiefly from the MS. of Froissart in the library of St. Elizabeth at Breslau in Prussia. Colonel Thomas Johnes was a Welsh squire, and at one time Lord Lieutenant of Cardiganshire : he established a private printing press at his residence, Hafod, where he issued his fine edition of Froissart, 1803-5. Colonel Johnes was also celebrated for his philanthropy, and especially for his zeal in forming plantations to cover the barren wastes of the district in which he lived. In four years, 1796-1800, he is said to have planted over two million trees. There is no evidence to prove that the great military crossbow of the fifteenth century, with its windlass, was in use at the time of Crécy (see remarks on crossbows at Crécy, pages 5, 6).

This crossbow, with its windlass, massive steel bow, stirrup, bow irons, wedges, long trigger and circular nut to hold the stretched string, shows no alteration from the time it was introduced, to the time when it was generally discarded in warfare, at the close of the fifteenth century.

The sporting crossbow employed for killing deer and other animals with an ordinary non-poisonous bolt, was, for about a hundred years, precisely the same in shape and mechanism—though rather smaller in size—as the large military weapon.

As this windlass crossbow was commonly used in sport and war for a long period, and as it is the one usually to be seen in museums and in mediæval illustrations, I will describe it in detail, as being the most interesting weapon of its kind to select for this purpose.

CHAPTER XIX

THE CONSTRUCTION OF A POWERFUL CROSSBOW, SUCH AS WAS USED IN THE FIFTEENTH CENTURY FOR KILLING DEER WITH A HEAVY NON-POISONOUS BOLT. THE SAME WEAPON, OF SLIGHTLY LARGER SIZE, WAS EMPLOYED IN WARFARE FROM ABOUT 1370 TO ABOUT 1490, OR TILL THE TIME WHEN MILITARY CROSSBOWS WERE GENERALLY DISCARDED FOR HAND-GUNS.

THE STOCK

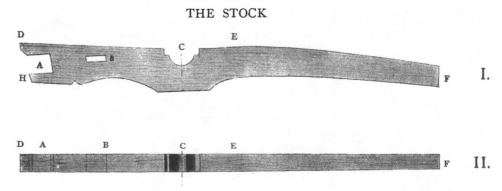

FIG. 46.—SIDE (I), AND SURFACE VIEW (II), OF THE WOODEN STOCK OF THE CROSSBOW, WITHOUT ANY OF ITS FITTINGS. Scale $\frac{1}{8}$ in. = 1 in.

A The opening to take the steel bow. The depth of this opening exactly fits the width of the bow at its centre, and is here 2 in. long and $1\frac{5}{8}$ in. deep.

The opening, it will be seen, is sloped upwards so as to give the bow the slight cant up, which, together with the upward curve of the ends of the bow, enables its string to act without friction along the groove on the top of the stock in which the bolt is laid.

———

B The oblong hole ($1\frac{3}{4}$ in. long, $\frac{1}{2}$ in. wide) into which the metal wedges (figs. 62, 63, p. 106) are driven which secure the bow tight to the stock.

The space between this hole and the opening for the bow at A, is 3 in.

c The hollow, cut transversely through the stock, in which the revolving nut and its socket are fitted, fig. 53, p. 97.

THE DIMENSIONS OF THE STOCK ARE:

Extreme length, D to F, 3 ft.

Depth at fore-end, D to H, $3\frac{1}{4}$ in.

Depth at small end F, $1\frac{3}{4}$ in.

Thickness, $1\frac{1}{2}$ in. from D to E, then tapering to $1\frac{1}{4}$ in. at the small end F.

From the point of the fore-end at D, to the centre of the opening at c (which is also the centre of the revolving nut when it is fitted, fig. 53, p. 97, 14 in.

The stock of a crossbow was always cut from some hard tough wood, such as beech, of close and straight grain ; the grain, of course, running lengthways with the stock to give it strength.

The sighting arrangement of a mediæval crossbow was of a rough and ready kind, though no doubt quick and effective in use. It consisted of a strip of wood of the same thickness as the stock, 1 ft. in length and $\frac{3}{4}$ in. high, fig. 47. The top of the strip was rounded, and had two or three large sloped transverse notches in it of varied depths, fig. 47.

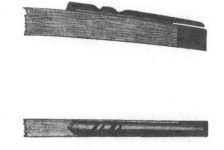

FIG. 47.—SIDE AND SURFACE VIEW OF THE SIGHT OF THE CROSSBOW.

Scale $\frac{1}{8}$ in. = 1 in.

This was screwed to the top of the small end of the stock.

The crossbowman grasped with his right hand the trigger and the handle-end of the stock of his crossbow, and took aim over the sharp point formed by the joint of the bent thumb, as it rested across one of the notches in the wooden strip. The first joint of the thumb and the uppermost edge of the head of the bolt, as the latter lay in the groove of the stock, gave the alignment. When the soldier was on the march with his crossbow over his shoulder, these notches bestowed a firm grip for the fingers of one hand.

The head of the bolt, whether blunt or pointed, being usually four-sided, had, therefore, four longitudinal edges. One of these edges was always arranged to be upright so as to act as a fore-sight, when the butt of the bolt was placed between the fingers of the nut and against the bow-string.

The notches in the strip of wood being of different depths, the thumb of the right hand, acting as a back-sight, could instantly be placed higher or lower, according to the trajectory required.

The after-end of the sighting strip (fig. 47, previous page), it will be seen, is cut away for a length of 3 in. and a depth of $\frac{1}{4}$ in. This was to allow the sheath of the windlass to be fitted over the end of the stock, fig. 73, p. 120 (upper plan).

The stock was covered at its end with a cap of thin metal for a length of 2 in., to protect it from the friction of the sheath of the windlass, A, fig. 47, previous page.

In the case of crossbows with long stocks, such as those bent with a windlass and its ropes, as here described, the small or pointed end of the stock (known as the tiller) was either squeezed tight inside the right arm-pit, or was rested for a few inches on the top of the right shoulder. The left hand grasped the enlarged part of the under surface of the stock, and the left elbow rested on the left hip or against the left side, in order to support the weapon in a horizontal position. The fingers of the right hand were thus free to work the trigger, and the right thumb to act as a back-sight. The face was inclined over the stock, so as to bring the right eye in line with the groove in which the bolt was laid, fig. 36, p. 78.

Louis XI. of France, 1461–1483, issued a military order that crossbowmen in his army should have the vizors of their helmets cut away on the right side opposite the cheek, so that the vizor might not interfere with the stock of the crossbow when the crossbowman was taking aim.

The sporting crossbows with short straight stocks, such as those bent with a cranequin (fig. 87, p. 135), were held just clear of the shoulder, those with enlarged butt-ends being placed against the top of the shoulder.

CHAPTER XX

THE CONSTRUCTION OF THE CROSSBOW (Continued)

THE REVOLVING NUT AND ITS SOCKET

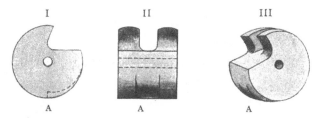

FIG. 48.—SIDE (I), FRONT (II), AND PERSPECTIVE (III), OF THE CIRCULAR STEEL, OR IVORY NUT WHICH HOLDS THE BOWSTRING WHEN THE BOW IS BENT.   Half full size.

THE notch A, in the nut, is exactly below—*i.e.* opposite to—the curved fingers which hold the bow-string.

The notch is $\frac{1}{2}$ in. wide, and $\frac{1}{8}$ in. deep on its squared face where it engages the point of the trigger inside the stock.

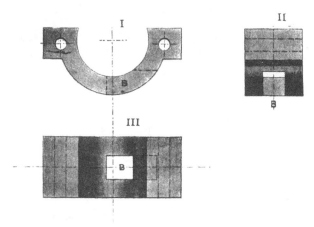

FIG. 49.—SIDE (I), END (II), AND SURFACE VIEW (III), OF THE METAL SOCKET IN WHICH THE NUT REVOLVES.   Half full size.

The longitudinal opening B ($\frac{1}{2}$ in. wide), is cut through the under side of the socket, to allow the point of the trigger to reach, and then engage in the notch in the nut, as shown in fig. 55, p. 98.

The nut and its socket should be of steel, and turned in a lathe to fit each other exactly, so that the nut may revolve accurately and closely in its

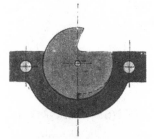

socket. The nut is $1\frac{1}{2}$ in. diameter and $1\frac{1}{4}$ in. thick.

In outline the nut is, of course, a circle. The socket in which it revolves, is $\frac{1}{4}$ in. more than a half circle, so as to bring the centre-hole of the nut $\frac{1}{4}$ in. below the surface of the socket, and also of the stock of the crossbow,[1] as shown in fig. 50.

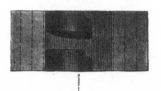

The pin ($\frac{1}{8}$ in. diameter) which passes through the $\frac{3}{16}$ in. hole in the centre of the nut, and also through the lock-plates, is merely intended, without receiving any pressure, to hold the nut in its position in the socket.

FIG. 50.—SIDE AND SURFACE VIEW OF THE REVOLVING NUT IN ITS SOCKET.

Half full size.

The socket should take all the pressure of the nut when the bow-string is stretched over the fingers of the latter, and for this reason the pin is slightly smaller than the hole in the centre of the nut. If any strain came upon the pin which passes through the nut, it would bend and the nut would not then revolve.

In many mediæval crossbows, the pin through the nut was omitted, though sometimes present in the form of a thin length of catgut passed several times through the hole in the nut, and then round the stock, just to prevent the nut from falling out of its socket and being lost, fig. 51.

More often, however, the nut, being only of horn, was not weakened by having a hole

FIG. 51.—A NUT SECURED BY CATGUT.

bored through its centre, but was held in its socket by two little screw-pins, one through each lock-plate, neither of which pins penetrated the opposite centres of the nut more than $\frac{1}{4}$ in., fig. 52.

---

[1] This prevents the pin which passes through the nut from being too near the upper edge of the stock. It also gives the revolving nut more 'centre bearing' against its socket to withstand the strain of the bow-string.

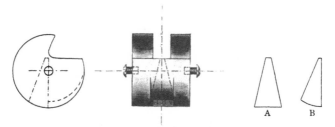

FIG. 52.—THE HORN NUT OF THE MEDIÆVAL CROSSBOW AND ITS STEEL WEDGE.
Half full size.

The nut and its socket were formerly both made of horn.[1] The nut was usually cut from the crown of a stag's antlers. This was a very tough material for the purpose, and also one that was light, and therefore free and quick in use and loose as applied to its connection with the bow-string.   In Scandinavia, however, walrus tusk was commonly used for the nut of a crossbow.

The horn nut always had its notch protected by a small wedge of hardened steel, which met the point of the trigger inside the stock. Fig. 52 shows this kind of nut, and A, B, the front and side view of its steel wedge separate from it.

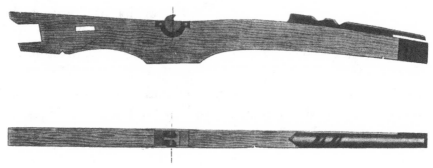

FIG. 53.—SIDE AND SURFACE VIEW OF THE STOCK OF THE CROSSBOW, WITH THE NUT AND ITS SOCKET IN POSITION.   Scale $\frac{1}{8}$ in. = 1 in.

The centre hole of the nut is 14 in. from the upper point of the fore-end of the stock.
(D—C, fig. 46, p. 92.)

---

[1] Steel nuts and sockets were not generally fitted to crossbows till about 1640-1650.

CHAPTER XXI

THE CONSTRUCTION OF THE CROSSBOW (Continued)

THE TRIGGER AND LOCK

Fig. 54. The handle or round part of the trigger, A–A, that is outside the stock, is $8\frac{1}{4}$ in. long and $\frac{3}{8}$ in. diameter. The flat part of the trigger, A B, which works inside the stock, is $\frac{7}{16}$ in. thick. The hole for the transverse pin on which the trigger hinges, is $\frac{3}{8}$ in. diameter.

The point, B, of the trigger, or that part of it which engages the notch in the nut, is hardened to withstand wear from friction.

FIG. 54.—SIDE VIEW OF THE TRIGGER OF THE CROSSBOW.

Scale $\frac{1}{4}$ in. = 1 in.

The point, B, of the trigger, is $\frac{7}{16}$ in. thick and $\frac{1}{4}$ in. deep. It should just fit through the opening in the socket, as well as for $\frac{1}{8}$ in., or half its depth, into the notch in the nut, as shown in fig. 55.

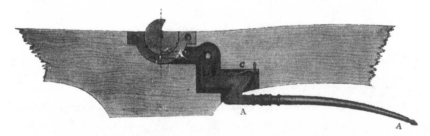

FIG. 55.—SIDE VIEW OF THE TRIGGER IN POSITION IN THE STOCK, SHOWING HOW THE LOCK OF THE CROSSBOW WORKS.

Scale $\frac{1}{4}$ in. = 1 in.

Fig. 55. When the handle end of the trigger, A–A, is pressed upwards towards the under side of the stock, the point of the trigger (B, fig. 54), at once drops out of the notch in the nut. The nut being then free to revolve, instantly releases the bow-string, which was stretched and previously held fast over its fingers.

The small spring (c, fig. 55), inside the stock, forces the point, B, of the trigger firmly into the notch of the nut. In this way the bow-string is securely held till the nut is released by pressing the handle of the trigger upwards.

After the fifteenth century, other forms of trigger were invented for holding and releasing the revolving nut. The lock here described was, however, the simplest and best for ordinary use, and till the end of the fifteenth century was the only one applied to crossbows, whether military or sporting, which discharged bolts.

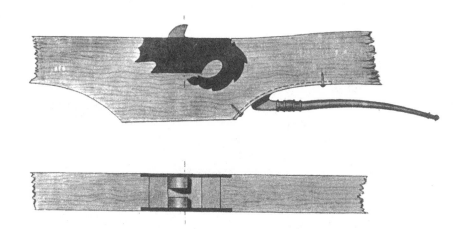

FIG. 56.—SIDE AND SURFACE VIEW OF THE NUT, SOCKET, TRIGGER, LOCK-PLATES AND TRIGGER-PLATE FITTED TO THE STOCK OF THE CROSSBOW.

Scale $\frac{1}{4}$ in = I in.

Fig. 56. The lock-plates—one on each side of the stock—are of steel, $\frac{1}{8}$ in. thick. The lock-plates and their transverse screws strengthen the stock where it is cut out for the nut and its socket; they also hold the nut, socket and trigger in position.

The lock-plates (shaded) are morticed in flush with the woodwork of the stock, and close against the sides of the revolving nut and its socket, fig. 56.

The trigger-plate is fitted beneath the stock, as per dotted line and screws.

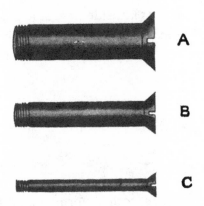

FIG. 57.—THE STEEL SCREW-PINS FOR THE LOCK-PLATES, THE PIN FOR THE TRIGGER AND THE ONE FOR THE REVOLVING NUT.

Full size.

A. The $\frac{3}{8}$ in. pin on which the trigger hinges.

B. The $\frac{1}{4}$ in. pins (5), which fasten the lock-plates, and also the socket for the nut.

C. The $\frac{1}{8}$ in. pin on which the nut revolves.

These pins all pass through the lock-plates and the stock, from side to side. They rigidly secure the lock of the crossbow to its stock.

When the pins are screwed into place, their heads and points should be level with the metal round them.

In mediæval crossbows, the pins of the lock were always riveted by a hammer at each of their ends after they were driven in. This was, perhaps, a tighter method of fixing them, but was a plan which prevented the lock from being readily taken apart.

CHAPTER XXII

THE CONSTRUCTION OF THE CROSSBOW (Continued)

THE STEEL BOW, THE BOW-IRONS AND THE STIRRUP

DIMENSIONS OF THE BOW (FIG. 58, NEXT PAGE).

LENGTH.—Between extremes, 2 ft. 6 in.

WIDTH.—At centre of length, $1\frac{5}{8}$ in., with a gradual reduction to a width of 1 in. at 2 in. from each end.

THICKNESS.—At centre of length, $\frac{1}{2}$ in., with a gradual reduction to $\frac{3}{8}$ in. at 2 in. from each end. Width across enlarged parts of ends, each $1\frac{1}{2}$ in.

The bow is flat on all sides, with squared edges.

For the ends of the bow, into the notches of which the loops of the bow-string fit, see fig. 69, p. 114.

The normal bend of the bow, taken from the centre of its length, inside its curve, to the centre of a thread connecting its ends, is $4\frac{1}{2}$ in., C—D, A, fig. 58, next page.

B, Fig. 58, next page, shows how the arms of the bow are slightly canted up from its centre. If a thread is held from the centre of one end of the bow to the centre of its other end, as per dotted line, it should be $\frac{1}{4}$ in. higher at its centre than the upper edge of the bow, as the latter lies on its side on a table, X—X, B, fig. 58.

If the bow had not this upward cant, its bow-string would press so hard on the top of the stock that it would be unable to propel the bolt with proper force. The friction of the bow-string against the stock would prevent the string from acting freely when the bow recoiled from a bent position. All the best steel bows were made in this manner, especially those used in the

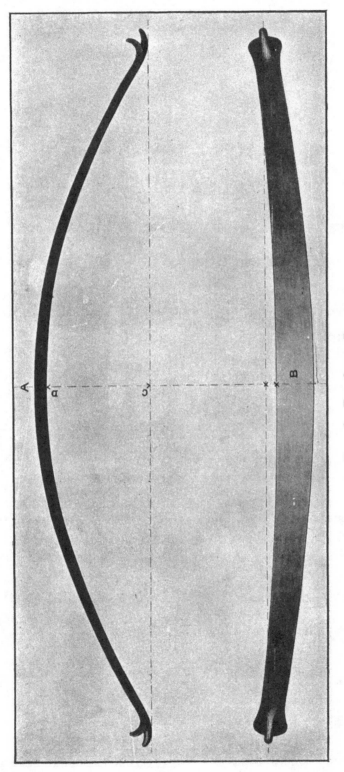

Fig. 58.—The Steel Bow.

A, Side View, showing the Normal Curve or Bend of the Bow.

B, Full-face View of the Front or Belly of Bow, showing how its ends are canted up from its centre.

Scale ¼ full size.

chase.   On the other hand, many of the military crossbows had straight bows, which were merely canted upwards in their stocks to enable their bow-strings to work freely.   The latter plan did not, however, give so straight a pull and so fair a strain to the bow, as the one described in fig. 58.

To procure a good bow of spring steel of correct size and shape, it should be first modelled in wood.   The model should then be sent to a spring maker to copy, with instructions to temper the steel a little soft, so that the bow may take a slight 'set,' rather than break, if overstrained.

Liège in Belgium is the best place from which to obtain a trustworthy bow.

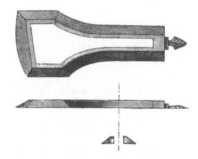

FIG. 59.—SURFACE, SIDE, AND SIDE-SECTION OF ONE OF THE BOW-IRONS.   Scale $\frac{1}{4}$ in. = 1 in.

There are two of these irons, one on either side of the stock.   They are each 7 in. long, $\frac{1}{4}$ in. thick, $\frac{1}{2}$ in. wide all round their sides and $\frac{1}{2}$ in. wide between the narrow parts of their openings.

The wide openings of the irons, at their large ends, exactly fit the width of the bow ($1\frac{5}{8}$ in.) at its centre.

The irons surround the centre of the bow, as well as the corners of the base of the stirrup.   The base of the stirrup rests upon the centre of the back of the bow.

The bow-irons act as straps to pull, and then immovably hold, the bow and its stirrup tight against the stock of the crossbow, this being achieved by the metal wedges presently described, figs. 61, 62, 63, p. 106.

When the bow, the bow-irons and the stirrup, are in position on the stock of the crossbow and ready for the wedges to be applied which secure them, the narrow ends of the openings in the bow-irons should each be $\frac{3}{4}$ in.

short of that end of the oblong hole in the stock which is next the nut, as shown at E, fig. 63, p. 106.

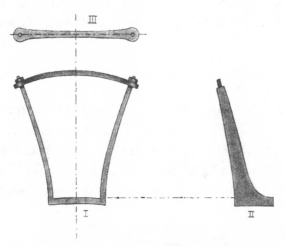

FIG. 60.—THE METAL STIRRUP: ITS FRONT (I), SIDE (II), AND TOP BAR (III).
Scale $\frac{1}{4}$ in. = 1 in.

The base of the stirrup fits close upon, and is the same width ($1\frac{5}{8}$ in.) as the centre of the back of the bow.

Its base is 2 in. long inside, or $\frac{1}{2}$ in. more than the thickness of the stock, I, fig. 60. This is necessary in order to give space for the bow-irons to encircle the corners of the stirrup, when the stirrup and the bow are placed in the opening in the fore-end of the stock, preparatory to their being fixed in position by the wedges acting on the bow-irons, figs. 61, 63, p. 106.

The crossbowman placed his foot in the stirrup, to enable him to hold his crossbow firmly to the ground, whilst he bent its bow with the wind-lass; or, in the case of small crossbows, as he drew the string to the nut with his hands alone or by means of a rope and pulley. See fig. 77, p. 124 for a crossbowman bending his steel bow with a windlass.

CHAPTER XXIII

THE CONSTRUCTION OF THE CROSSBOW (*Continued*)

HOW TO FIX THE BOW TO THE STOCK

FIRST secure the stock of the crossbow perpendicularly in a vice, its fore-end upwards.

Take the stirrup and bow-irons together, as shown in fig. 61, next page, and pass the bow through the irons, the centre of the back of the bow being against the base of the stirrup. Place the base of the stirrup and the bow in the opening (A, fig. 46, p. 92), in the fore-end of the stock, with a bow-iron on each side of the stock.

Insert the short guard, A, into the oblong hole in the stock, at the end of the hole next the bow, figs. 62, 63, next page.

The angled ends of this short guard fit over the wood of the stock between the sides of the bow-irons.

Next insert the long guard, B, through the bow-irons, against their narrow ends, and through the oblong hole in the stock, figs. 62, 63, next page. The angled ends of B, turn back over the top of the solid part of the narrow ends of the bow-irons, and in this way they hold the latter close against the stock. Now push in between the guards, A, B, the two wedges, C, D, from opposite sides of the stock, figs. 62, 63, next page. By hammering in these wedges, the bow-irons will gradually draw the base of the stirrup, and hence the centre of the bow beneath it, with great force tight against the stock.

The empty $\frac{1}{2}$ in. space (E, fig. 63, next page) of the oblong hole in the stock, is left in case further tightening up of the bow is ever necessary. This tightening can easily be done by fitting a thin metal strip, to act as a washer, at the back of one of the guards.

Before finally fixing the bow to the stock, make certain that its position is correct. If the bow is to shoot accurately and with its full power, three matters require careful attention.

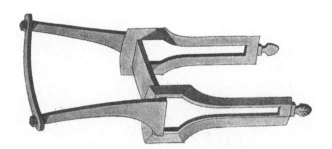

FIG. 61.—THE STIRRUP AND THE BOW-IRONS, READY TO TAKE THE BOW AND TO BE FITTED WITH THE BOW TO THE OPENING IN THE FORE-END OF THE STOCK.

Scale $\frac{1}{4}$ in. = 1 in.

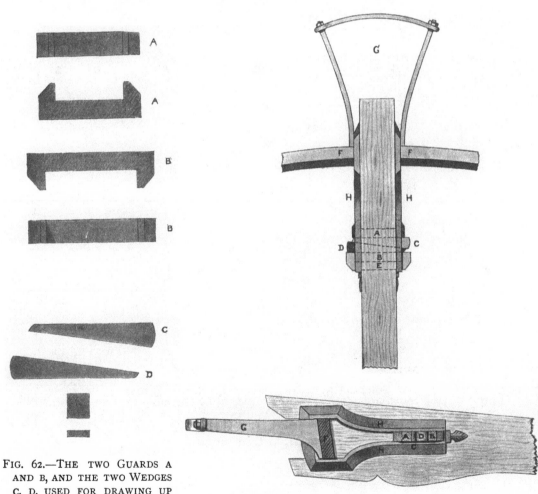

FIG. 62.—THE TWO GUARDS A AND B, AND THE TWO WEDGES C, D, USED FOR DRAWING UP THE BOW-IRONS WHICH FIX THE BOW TO THE STOCK. THESE ARE OF WROUGHT IRON.

Half full size.

FIG. 63.—FRONT AND SIDE VIEW OF STIRRUP G, BOW F, BOW-IRONS H, H, GUARDS A, B, AND WEDGES C, D, FIXED IN THEIR PLACES IN THE STOCK OF THE CROSSBOW.

Scale $\frac{1}{4}$ in. = 1 in.

I. The bow must be immovably fixed in the stock.

II. The arm of the bow on one side of the stock, should not be even $\frac{1}{8}$ in. longer or higher than its arm on the other side of the stock. A piece of twine tightly fastened from one end of the bow to the other, with a little bit of coloured silk knotted round the exact centre of its length, will be a guide of much assistance when regulating the position of the bow.

III. When the bow is fixed, a thread of cotton stretched from the centre of one end of the bow to the other, should be in a straight line and not pushed up at its centre by the surface of the stock.

The thread should cross $\frac{1}{4}$ in. above the stock, so that when the bow-string, which is $\frac{1}{2}$ in. in diameter, is fitted to the bow, the lower edge of the string will just lightly touch the groove in which the bolt is laid. There will then be no friction to retard the force of the bow-string when the crossbow is discharged.

CHAPTER XXIV

THE CONSTRUCTION OF THE CROSSBOW (Continued)

THE GROOVE FOR THE BOLT

FIG. 64.—THE BRASS GROOVE FITTED INTO THE SURFACE OF THE STOCK.

Scale ¼ in. = 1 in.

IN this groove the bolt is laid when the crossbow is ready for discharging.

The groove ($13\frac{1}{4}$ in. long, $\frac{3}{32}$ in. deep in its hollow) may be of brass. It reaches from the fore-end of the stock of the crossbow to the metal socket which holds the revolving nut, fig. 64.

The short ($\frac{3}{4}$ in.) length of that part of the metal socket which is in front of the nut, is recessed to correspond with the long separate grooved piece which comes up to it, fig. 64.

The groove should be neatly and tightly morticed in flush with the surface of the stock, fig. 65. It should have two or three thin pins to secure it from slipping forward, and will have to be as smooth as glass and as true as a gun-barrel from end to end.

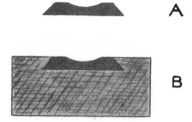

FIG. 65.—A, End Section of the Brass Groove. B, End Section of the Fore-end
of the Top of the Stock, with the Groove driven into its Mortice.

Full size.

In many old crossbows the groove for the bolt was of horn, and glued into
its mortice on the top of the stock. This was a lighter method, and is one to
be recommended if a suitable piece of horn is available.

CHAPTER XXV

THE CONSTRUCTION OF THE CROSSBOW (Continued)

THE BOW-STRING

THE bow-string should be composed of several dozen turns of thin twine, of pure hemp or flax. What is known in seaports as 'sailmakers sewing twine' is excellent for the strings of large crossbows, as it is very strong and will not stretch under the great strain of the steel bow. Any twine in the form of soft twisted string is sure to stretch, and what was at first a taut bow-string will—if this kind of material is used—soon become a slack and useless one.

For small crossbows, such as those presently described for target practice and rook shooting, there is no twine so suitable for their bow-strings as that employed by bookbinders for stitching the leaves of a book to its back.

Bookbinder's thread is extremely strong and hard, and though it will not stretch, it is rather too fine for the string of a large crossbow.

The string of a crossbow should always be taut, so that when the weapon is discharged, the bolt receives sufficient impetus at the moment the bow-string is checked by the ends of the bow, as the latter straightens.

In some foreign longbows, an arrow can be driven as far with a loose bow-string as with the more usual tight one, but in a crossbow the draw of the string along the stock is so short (from 5 in. to 6 in. only), that all the power of the bow has to be utilised in this small length of pull.

In mediæval pictures, the string of the crossbow is often represented with the same thickness at its looped ends as at its centre. This suggests that in old days, the string was taken its full thickness round each extremity of the bow, and that the ends of the string were then in some way knotted or wrapped, to secure them and form the necessary loops.

In what manner this could have been done, I have never been able to ascertain.   I find that the great strain put upon the bow-string of a crossbow by its lever or its windlass, forces any modern knot or wrapping to draw, with the result that the string soon becomes slack, though perhaps tight enough when first fitted to the bow.

Splicing is not feasible, because the string of a crossbow is not laid in coils like a rope, but consists of many fine threads stretched straight from one end of the bow to the other.   If the bow-string was made in twisted strands like a rope, it would at once stretch and become useless.

Nor is the timber-hitch fastening (as is used to form the loops of the string of the longbow) possible, owing to the great bulk of the string of the crossbow.

It is evident, however, that if the bow-string of a crossbow could be made with its end loops of the same thickness as the rest of the bow-string, the string would be stronger than the one I am about to describe, in which the turned parts of the string that form the loops, are only half the substance of its centre.[1]

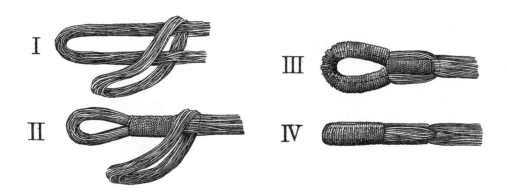

FIG. 66.—HOW EACH END OF A CROSSBOW STRING WAS SOMETIMES STRENGTHENED BY AN AUXILIARY LOOP.

I., II. An auxiliary loop of fine thread passed between the halves of the bow-string.

III., IV. The two loops lashed together so as to jointly form one end of the bow-string.

In some mediæval crossbows, the bow-strings were strengthened at their ends in a very ingenious manner by means of auxiliary loops.   How this was done is shown in fig. 66.

[1] I have, however, never known a string to break if made as here directed.

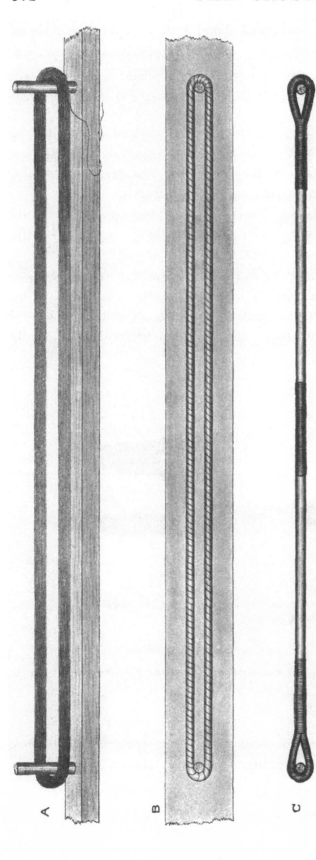

FIG. 67.—A, THE SKEIN. B, THE SKEIN WRAPPED WITH FINE THREAD. C, THE FINISHED BOW-STRING.

HOW TO MAKE THE BOW-STRING OF A CROSSBOW WHICH SHOOTS BOLTS, FIG. 67

Hammer a round peg of hard wood (4 in. long, $\frac{1}{2}$ in. diameter) firmly into a hole drilled through a board which is 3 ft. in length, by 6 in. wide and 1 in. thick. The hole for the peg should be 3 in. from one end of the board, and the peg should be set perfectly upright.

Place the notch at one end of the steel bow level with this peg, then fix a second ·peg in the board $\frac{1}{2}$ in. short of the notch at the other end of the bow. The measurements should be taken from the outside edges of the pegs.   This will give you the correct length of the bow-string.

Next wind the fine twine evenly round and round the two pegs in the board—being careful not to cross the threads between the pegs—till you have a smooth tight skein (A, fig. 67), which—when its halves are wrapped together for an inch or so with a little piece of twine to test its diameter—is the thickness of the small joint of your little finger, or $\frac{1}{2}$ in.

Now rub beeswax all over the skein till its threads stick together. This will make the bow-string impervious to water or damp and will preserve it indefinitely.

_____

Without removing the skein from the pegs, wrap a long well-waxed length of strong silk—in turns $\frac{1}{8}$ in. apart—round its entire length, and a little closer (by the aid of a darning-needle) at its ends where they pass round the pegs, B, fig. 67.

Without this wrapping, the skein is sure to fall (especially at its ends), into a hopeless tangle during the process of converting it into the bow-string.

_____

With some hard twine—about the substance of an ordinary knitting-needle—tightly wrap the skein (or bow-string as it may now be called), as shown in C, fig. 67.

The centre wrapping, which lies above the groove in the stock, is 4 in. long, and the end wrappings are each 3 in. long. The centre wrapping may be overlaid at its centre, for $\frac{1}{4}$ in., with a little crimson silk, to show the exact centre of the bow-string, so that should the bow shift a trifle in use, the movement can be at once detected and the bow re-adjusted in the stock.

_____

The loops at each end of the bow-string should be wrapped, if possible, without removing the skein from the pegs in the board. If this cannot be achieved, on no account lift the skein off the pegs till the centre and end wrappings are completed.

When the bow-string is finished, the silk can be removed where it shows between the wrappings.

CHAPTER XXVI

THE CONSTRUCTION OF THE CROSSBOW (Continued)

HOW TO FIT THE BOW-STRING TO THE BOW

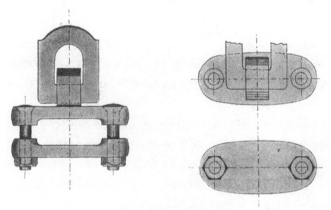

FIG. 68.—ONE OF THE METAL CLAMPS TO WHICH THE BASTARD STRING IS ATTACHED. FRONT AND SURFACE VIEWS.

Half full size.

THE bow-string is, as we know, $\frac{1}{2}$ in. shorter than the space between the notches of the bow, hence it will not reach from one notch of the bow to the other, when held between them. Of course the bow-string would not be taut when the bow is strung, if it were not shorter than the bow.

To place the loops of the bow-string over the ends of the bow, in the notches shaped to receive them, mechanical aid is necessary. It would be impossible to bend a thick steel bow enough for this purpose by manual power alone.

To fit the bow-string to a crossbow, what was termed a false or bastard string was employed.[1] The bastard string—by means of the windlass of the crossbow (fig. 76, p. 123)—bent the steel bow sufficiently to allow the loops of the bow-string to be slipped over the ends of the bow into its notches, as shown in fig. 70.

FIG. 69.—ONE OF THE CLAMPS SCREWED TO ONE END OF THE BOW, WITH ONE END OF THE BASTARD STRING ATTACHED TO IT. SIDE VIEW.

Half full size.

[1] 'A spindle full of raw thread to make a false string for the king's crossbow.'—*Monstrelet's 'Chronicles.'*

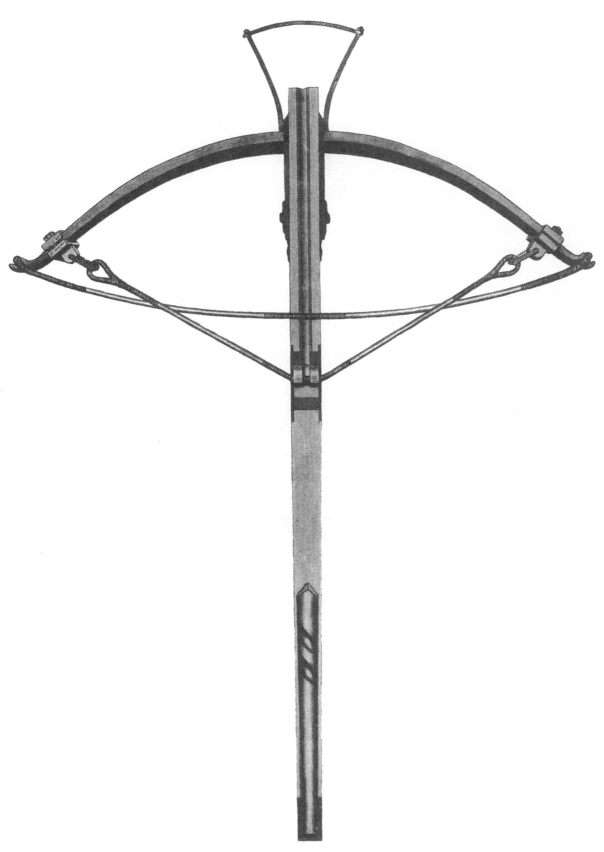

FIG. 70.—THE CROSSBOW WITH ITS BOW SUFFICIENTLY BENT BY THE BASTARD STRING TO ALLOW ITS BOW-STRING TO BE FITTED INTO THE NOTCHES AT THE ENDS OF THE BOW, WHICH IS HERE DONE.

The bastard string was then removed from the bow till next required, fig. 71.

The bastard string (in its construction similar to the bow-string) was temporarily fixed to the arms of the steel bow by two little iron screw clamps, fig. 70, previous page. It hung rather loosely between the clamps when the latter were attached near the ends of the bow. The windlass was then used to pull the bastard string tight down over the fingers of the nut, where it was held fast whilst the bow-string was being fitted, fig. 70.

By regulating the position of the clamps on its bow, any crossbow could be bent by its windlass just enough to enable its bow-string to be removed or replaced.

To remove the bastard string (after having fitted the bowstring into the notches of the bow), do not pull the trigger of the crossbow. Hold one handle of the windlass with one hand and press the trigger at the same time with your other hand, then let the bastard string gradually slacken and the bowstring tighten as you reverse the windlass.

The fitted bow-string should be from $\frac{1}{2}$ in. to $\frac{3}{4}$ in. further along the groove in the stock, towards the nut, than a thread would be, if stretched between the ends of the bow whilst the latter was at rest previous to its bow-string being put on. In this crossbow the front of the string should be 5 in. from the inside upper edge of the centre of the bow, and the back of the string, 6 in. from the centre of the nut.

If the fitted bow-string is a little slack, take it off the bow by means of the bastard string. Undo the centre wrapping, give the string two or three twists to shorten it, then replace it on the bow and wrap its centre again.

If the string is very tight, and bends the bow too much—the string being, for instance, $1\frac{1}{2}$ in. along the stock instead of from $\frac{1}{2}$ in. to $\frac{3}{4}$ in. as above explained—there will be a waste of power; also a risk of fracture when the bow is fully bent by the windlass.

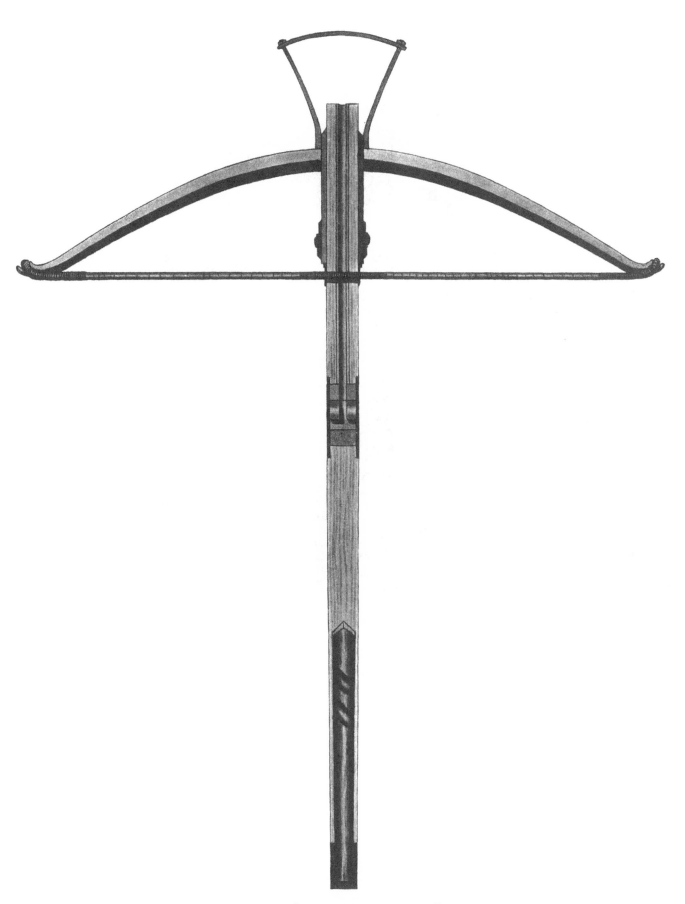

FIG. 71.—THE CLAMPS AND THE BASTARD STRING REMOVED AND THE BOW FITTED WITH ITS BOW-STRING.

Fig. 72.—Crossbow finished.

A string which is $\frac{1}{2}$ in. too long, can always be set right, but a string which is $\frac{1}{2}$ in. too short is beyond remedy. The only method in the latter case, is to unwind the skein of which the string is composed and remake it longer.

The crossbow is now completed, with the exception of its windlass. It should appear as in fig. 72, with its woodwork smooth and nicely stained, and its metal fittings hardened and polished. The sharp edges of the stock can all be rounded off slightly, saving, of course, the opening in which the bow and its stirrup are fitted.

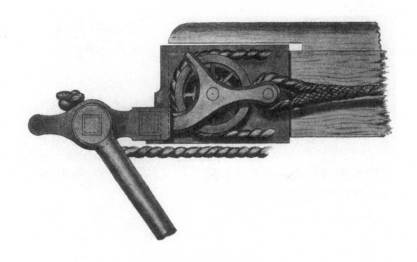

FIG. 73.—THE HANDLE END OF THE WINDLASS, SURFACE AND SIDE VIEW.
Half full size.

CHAPTER XXVII

THE CONSTRUCTION OF THE CROSSBOW (Continued)

THE WINDLASS

THE sheath of the handle end of the windlass fits over the small end of the stock of the crossbow, as shown opposite in fig. 73. In the surface view, only the ends of the cords are inserted, to avoid confusion of detail. The end of the sighting strip is also omitted in this view for the same reason.

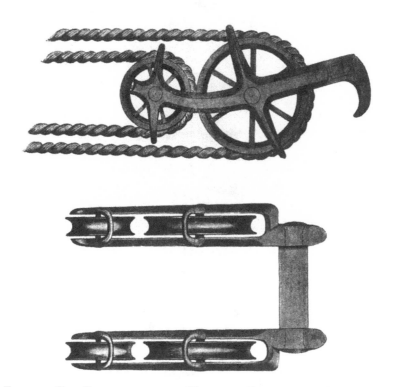

FIG. 74.—THE FORE-END OF THE WINDLASS, SURFACE AND SIDE VIEW.
Half full size.

Fig. 74. In the surface view, the pulley cords are not given, in order to show more clearly the arrangement of the wheels and of the guards which keep the cord in position on the wheels. See fig. 75, next page, for the windlass in position on the crossbow.

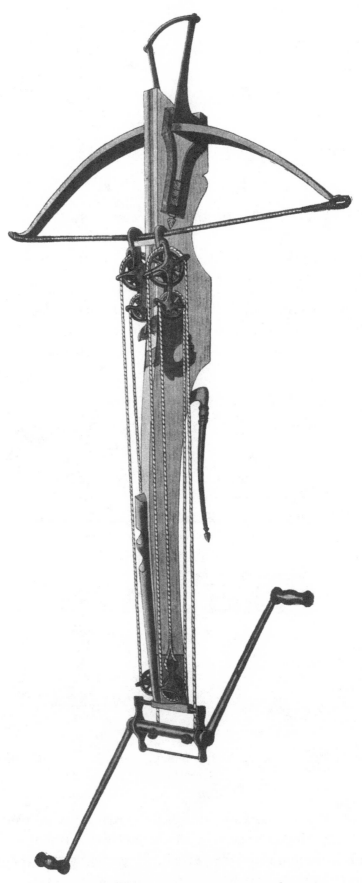

FIG. 75.—THE WINDLASS ATTACHED TO THE CROSSBOW,
PREPARATORY TO BENDING ITS BOW.

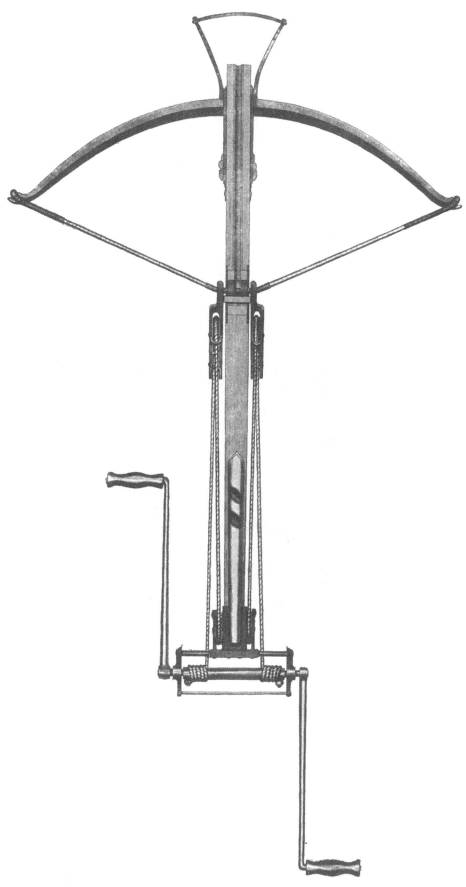

FIG. 76.—THE CROSSBOW WITH ITS BOW BENT BY THE WINDLASS AND ITS
BOW-STRING SECURED OVER THE FINGERS OF THE NUT.

To draw the bow-string of a powerful crossbow to the nut, a windlass
or a cranequin is necessary. Though the distance which the bow-string has
to be pulled along the top of the stock is only some 5 or 6 in., no manual
strength could draw it half-way.

FIG. 77.—CROSSBOWMEN—FIFTEENTH CENTURY.

The stooping figure has a windlass crossbow, and is winding up the bowstring of his weapon.
The erect figure carries a crossbow that is bent by the metal claw which may be seen hanging
from his belt.   See Chapter XV. for a description of the belt and claw.

From C. Leberthais' 'Ancient Tapestries of the City of Rheims.' Paris, 1843.

A crossbow windlass, small though it be, has immense power, and will
draw the bow-string to the nut smoothly and quickly, and with no perceptible
strain or exertion.

To use the windlass, the sheath of its handle end is fitted over the small end of the stock, and the claws of its fore-end are hooked over the upper surface of the bow-string, as shown in fig. 75, p. 122.

Fig. 76, p. 123.   In this plan, though the bow is bent, the windlass is not removed, so that its proper position on the stock and string may be seen.

By reversing the handles of the windlass a couple of turns to slacken its cords, it can be quickly taken off the stock of the crossbow, which is then ready for use.   In former days the crossbowman, after he had stretched his bow-string and removed the windlass, suspended the latter from his side, by means of a hook attached to his belt.

Fig. 77 shows a crossbowman using his windlass to bend his steel bow.

He holds (as was usual) a bolt between his teeth that it may be ready at hand to place on the stock of his crossbow when its bow is bent.   See fig. 19, p. 49, for another example of this practice.

CHAPTER XXVIII

THE CONSTRUCTION OF THE CROSSBOW (Concluded)

THE BOLT, OR QUARREL, AND HOW IT WAS ARRANGED ON THE STOCK OF THE CROSSBOW

FIG. 78.—THE BOLT FOR THE CROSSBOW (OF ASH OR YEW)

DIMENSIONS

Total length	$12\frac{1}{2}$ inches.
Length of head	3 ,,
Diameter of shaft at c, near where it meets the sheath of the metal head	$\frac{11}{16}$ of an inch.
The height of the shaft at its butt-end . .	$\frac{1}{2}$,, ,,

The weight of the bolt is $2\frac{1}{2}$ oz. (shaft 1 oz., metal head $1\frac{1}{2}$ oz.).

The butt of the shaft, for about an inch, A—A, is slightly flattened at opposite sides (in line with the side feathers) and then tapered to a width of $\frac{3}{8}$ in., figs. 78, 79. This allows the butt to fit between the fingers of the nut and against the bow-string, as shown in fig. 82, p. 128.

The butt of the shaft, being tapered as well as flattened, can be gently wedged in between the fingers of the nut to prevent the bolt from slipping forward when the crossbow is aimed towards the ground.

FIG. 79. —END VIEW OF THE BUTT OF THE BOLT AND THE FEATHERS.

Full size.

It will be seen that the head of the bolt is of greater width than height, fig. 80.

The head was thus shaped so that it might not touch the groove of the stock, and cause friction to or divert the direction of the bolt when it was propelled by the bow.

FIG. 80.—SECTION OF THE METAL HEAD OF THE BOLT.

Full size.

The bolt should lie with only the enlarged fore-end of its wooden shaft (c, fig. 78), and its butt resting upon the groove of the crossbow, for reasons given p. 70.

The side edges of the head of the bolt should be in line with the side feathers of the shaft, and the upper edge of its head,—as the bolt lies in the groove on the top of the stock,—in line with the top feather of the shaft. One edge of the head of the bolt will then always be upright, to act as a long and fine fore-sight.

Previously to bending his crossbow, the crossbowman revolved the fingers of the nut downwards into its metal socket, in the direction of the bow, fig. 81. When the bow-string was drawn along the groove of the stock by the windlass, it finally encountered the flat surface of the nut at A, and pushed the nut round into the position which caused its notch to engage the point of the trigger inside the stock.

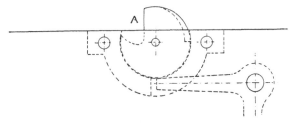

FIG. 81.—SHOWS HOW THE CROSSBOWMAN PLACED THE REVOLVING NUT BEFORE BENDING HIS CROSSBOW.

Half full size.

The fingers of the nut being then turned upright again, and being in front of the bow-string, and the notch in the nut having interlocked with the pointed end of the trigger, the stretched string was automatically caught, and thus held fast till the trigger was pressed to discharge the crossbow.

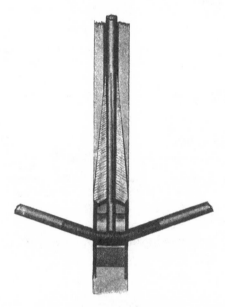

Fig. 82 shows how the stretched bow-string is held by the fingers of the nut when the bow is bent, and how the bolt is placed on the groove of the stock, with its butt-end between the fingers of the nut and close against the bow-string.

We also see in fig. 82 how the butt-end of the bolt is slightly tapered, so that it may be gently wedged in between the fingers of the nut, in order that it may retain its position should the crossbow be aimed downwards or carried over the shoulder.

FIG. 82.—THE BOWSTRING ON THE NUT
AND THE BOLT IN POSITION.
Scale ¼ in. = 1 in.

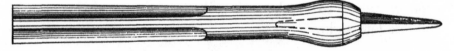

FIG. 83.—CROSSBOW BOLT WITH FLANGES CUT IN ITS SHAFT TO TAKE THE PLACE
OF FEATHERS.   Length 7 in., diameter of shaft ½ in.

Fig. 83. This curious variety of bolt was shown me by Col. Henry Walrond, the noted authority on Archery. I find that bolts made in this manner fly with great accuracy and force. They were, however, intended for the target and not for warfare, for which purpose a longer and heavier missile would be necessary. In the latter case, the flanges would require to be of a size that would prevent the bolt being used in a crossbow.

CHAPTER XXIX

THE SLURBOW

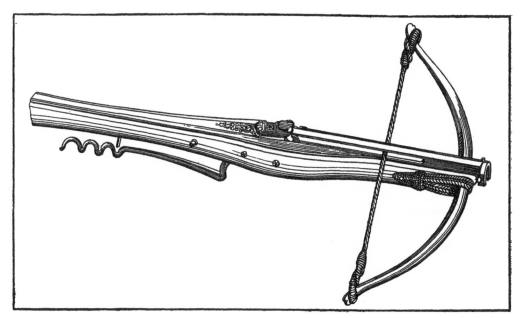

FIG. 84.—THE SLURBOW.

THE slurbow was a crossbow with a barrel of wood or metal attached to its stock. The barrel was no doubt suggested by the hand-gun. In the slurbow, the barrel was cut away to allow the bow-string to slide along the stock, as shown in fig. 84. See fig. 159, p. 220, for a modern slurbow.

The barrel sometimes consisted merely of a piece of wood, which was fitted to the crossbow above its grooved surface. The inside of this piece of wood was hollowed out to match the groove of the stock, so that the two together formed a tube in which to place the bolt.

I can find no good example of the mediæval slurbow except the one shown in fig. 84, which is dated 1549 and is in the Royal Armoury at Madrid.

These weapons are not mentioned till nearly the end of the fifteenth century, though in the first quarter of the sixteenth century, 'slurbows and their arrows,' 'slurbow bolts,' and 'fire arrows for slurbows,' are frequently referred to in lists of weapons stored in armouries and castles.

It is evident that the slurbow was used in warfare to discharge bolts, and not bullets.   The bolt used with a slurbow had no feathers on its shaft, as a feathered bolt could not have been placed inside the barrel of the weapon, see No. 7, fig. 10, p. 18.   Whether the slurbow was employed for sporting as well as for military purposes, I cannot say.   Its steel bow was bent by a cranequin.

CHAPTER XXX

THE SIXTEENTH-CENTURY SPORTING CROSSBOW, WITH A THICK STEEL BOW, WHICH WAS BENT BY A CRANEQUIN

ARBALESTE A CRIC—ARBALESTE A CRANEQUIN—RATCHET CROSSBOW— RACK AND PINION CROSSBOW

IN the last quarter of the fifteenth century the crossbow with its ropes and windlass, and its stirrup and necessarily long and heavy stock (to suit the action of the windlass), fell into disfavour with hunters of deer and other animals. The reason of this was the gradual perfecting of an instrument for stretching the string of a crossbow which enabled the crossbowman to dispense with the cumbersome metal foot-stirrup, and to have a much shorter and lighter stock to his weapon than had hitherto been possible. This apparatus was the French 'cranequin' or ratchet-winder. Not only was the cranequin simpler in use and more portable than the windlass, but it was also nearly as powerful, and, as it had no cords to fall out of repair, was practically indestructible.[1]

The cranequin was, however, considerably slower to work than a windlass, and could not, for instance, wind up the bow-string of a military crossbow with the rapidity necessary in open warfare. Its costly construction was also against its adoption by soldiery.

Slow to manipulate and expensive to make in comparison with the windlass, I doubt if the cranequin crossbow was ever popular in armies, though it is quite likely that it was employed by picked marksmen from behind the loop-holes and battlements of a besieged fortress. It is stated by various authors on military subjects, that in the fifteenth century crossbowmen were sometimes known as 'crenequiniers,' from their practice of shooting at the besiegers of a castle or town through the crenelles (loop-holes) in its walls.[2]

---

[1] The mechanism of the cranequin was precisely the same as that of the old-fashioned lifting jack to be seen in timber yards.

[2] Crenequin or crennequin is a way of spelling this word only to be found in modern works on armour and weapons. Cranequin is the old French spelling, and for this reason I consider there is no foundation for the surmise that the word 'cranequin' is in any way derived from 'crenelle.' In manuscripts of the fifteenth and sixteenth centuries, cranequin is sometimes spelt crannequin, carnequin, carnequyn and carnequing. Littré, in his *Dictionnaire de la Langue Française*, states that 'cranequin is derived from the Low German, "Kraeneke, a crane," owing to its shape.' The crossbow was itself often called a 'cranequin,' and the crossbowman a 'cranequinier' or 'crannequinier.'

In illustrations of the fifteenth and sixteenth centuries, mounted men are shown with these crossbows, fig. 16, p. 36, and fig. 18, p. 47.   As the cranequin

FIG. 85.—CROSSBOWMEN.

Reduced from an oil painting representing the martyrdom of St. Sebastian, its date 1514. The original is in the Church of St. Elizabeth at Marburg in Prussia.

The figure on the left is winding up his crossbow with a cranequin, the manner of doing this being very accurately shown.   The right-hand figure has his cranequin crossbow bent, and is placing a bolt in position.

From ' Costumes of Mediæval Christendom,' Hefner Alteneck, 1840–1854.

was the only crossbow winder which could be used on horseback to bend a strong steel bow, it is quite possible that mounted troops of superior rank

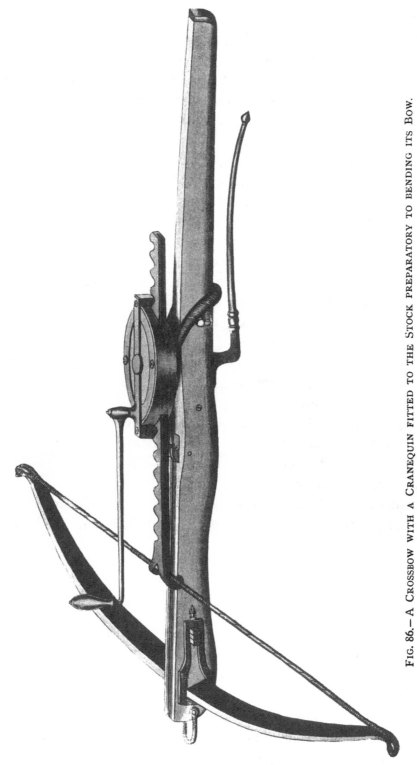

FIG. 86.— A Crossbow with a Cranequin fitted to the Stock preparatory to bending its Bow.

carried cranequin crossbows in action, as, for example, the select bodyguard of crossbowmen who guarded the person of Francis I. at Marignano in 1515.

Regarding the rate of movement of a cranequin, as compared with the windlass which preceded it, a cranequin with the usual handle of 9 in. in length, requires its handle knob to be turned round in a complete circle thirty times to draw back the bow-string of a crossbow $5\frac{1}{2}$ in., which was the common distance for the string to travel along the stock to the catch of the lock. This entails the manipulator moving his hand a space of 140 feet, and occupies him thirty-five seconds at a fair speed.

With a windlass, I find that the bow-string of the same crossbow can be drawn to the catch of its lock in twelve seconds; also that bolts can be discharged from a windlass crossbow at the rate of one a minute, whilst with a cranequin crossbow, the rate of discharge is two bolts in three minutes. Anyhow, nearly all the best sporting crossbows made after about 1480 and intended for killing deer, were fitted with 'cranequins' as winders for their bow-strings instead of with windlasses.

For sporting crossbows this winder was admirably adapted, and in their case speed in action was of no great consequence. The cranequin crossbow may be known by the increasing width of its short stock near the lock, and by the transverse iron pin which projects an inch or so on each side of the stock about seven inches behind the catch which holds the stretched bow-string. Against this pin the cord loop of the cranequin was rested, preparatory to using the latter to bend the bow, figs. 86, 87.

In crossbows bent by a goat's-foot lever, the transverse pin for the fork of the lever to rest on was fixed through the stock just below the catch of the lock, fig. 45, p. 89.

I can find no cranequin or even an illustration of one of a date previous to 1480, though I know of several crossbows made about 1460 that have the projecting metal pins through their stocks which indicate that cranequins were applied to bend their bows.[1]

The earliest cranequins to be seen in Continental and other armouries, date from about 1480. From this period the cranequin shows no change in its mechanism for some 150 years, or till the time when the crossbow with a heavy steel bow was no longer used for sporting purposes. The cranequin was a clever contrivance and acted perfectly, as it was able to stretch the

----

[1] It is probable that the cranequin was invented about the end of the first half of the fifteenth century.

string of a strong steel bow smoothly and easily, and with but slight exertion on the part of the operator.

Nearly all the cranequins I have examined are lavishly decorated, a proof that they were formerly owned by sportsmen and not by soldiers.

Between 1480 and 1530, I can find little variation in the construction of the powerful bolt-shooting sporting crossbow which was bent by a cranequin. About 1540, this weapon was, however, much improved. Its hitherto long and pointed stock was shortened and was also made with an enlarged butt-end to place against the shoulder.

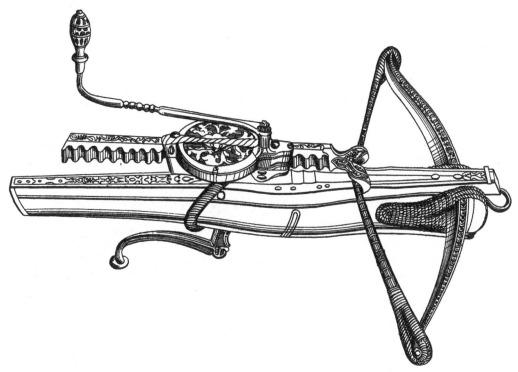

FIG. 87.—DECORATED CROSSBOW AND ITS CRANEQUIN, THE LATTER BEING IN POSITION FOR BENDING THE BOW. German, sixteenth century.

Though this new form of butt could not be sloped downwards like that of an arquebus, the alteration was a great convenience in aiming, especially as the side of the butt was hollowed out to receive the right cheek so that the right eye might glance along the top of the stock when aim was taken.

The long tapering stock of the crossbow bent by a windlass and its ropes, could not be fitted with an enlarged butt, as the small casing or box of the windlass had to fit over the pointed end of the stock, fig. 75, p. 122.

An easy pulling short trigger like that of a modern gun, together with a trigger-guard, was also fitted to the stock of the cranequin crossbow, in place of the long unprotected lever which previously, and from time immemorial had acted as a trigger. The old-fashioned long trigger required considerable pressure, with a consequent unsteadiness of aim, to cause it to discharge the crossbow.

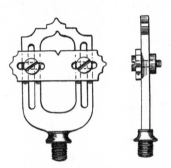

FIG. 88.—THE BACK-SIGHT OF A SIXTEENTH-CENTURY SPORTING CROSSBOW.

The crossbar of the back-sight could be moved up or down, or to the right or left, and then fixed by the small screws, when its notch and the head of the bolt were in a correct line for aiming.

In a cranequin crossbow, the back-sight was hinged to the stock, so that it might fold down flat when the cranequin was being used to bend the bow.

Besides the new style of trigger (suggested doubtless by that of the arquebus), the crossbow was fitted with a second or safety trigger.[1] This safety trigger prevented accidents, especially on horseback, as the crossbow could not be shot off till the safety trigger was pulled back to allow the front trigger to act, Chapter XXXVI.

Finally a back-sight was added to the stock of the crossbow to complete the weapon, fig. 88. The upper edge of the head of the bolt (as it lay on the surface of the stock) being viewed through the notch in the crossbar of the back-sight, gave the sportsman his alignment when taking aim, fig. 88.

A curious feature in most of the improved sporting crossbows of the sixteenth century bent by cranequins, is the absence of a groove along the top of the stock in which to lay the bolt preparatory to its discharge.

In fig. 89, opposite page, we see how the bolt of this form of crossbow was placed on the stock of the weapon.

The crossbow, it will be noticed, is flat on its upper surface at the part where in the older weapons there was a groove for the bolt.

In this case, the bolt rested near its head on a small raised cross-piece of ivory, which was fixed across the stock of the crossbow close to its fore-

---

[1] Nor was the hand-gunner above taking a hint from the crossbowman. In the earlier hand-gun, the bullet rattled loosely down the barrel; hence the chief cause of its failure in regard to range and accuracy was 'windage,' or the escape of the charge of powder, when ignited, past the sides of the bullet. The hand-gunner then bethought him of the crossbow bolt instead of a bullet, and found that he could shoot with great force a heavy-headed featherless bolt which exactly fitted the barrel of his hand-gun. For some years, bolts like crossbow bolts, called musquet-arrows, but without, of course, feathers, were frequently fired from hand-guns in warfare, both on land and sea.

Chroniclers tell us that these bolts, as discharged from hand-guns, were propelled with such power that they pierced from side to side the bulwarks of ships.

end. This little piece of ivory had a small hollow in its centre for the shaft of the bolt to lie in.

The butt of the bolt was placed against the bow-string, between the fingers of the revolving nut. The bolt being slightly tapered at its butt, the fingers of the nut held it from movement when it was gently wedged between them. It will be observed that the shaft of the bolt—except at its extremities—does not come in contact with the crossbow. With this plan of arranging the bolt, friction was much reduced and the missile left the crossbow with an accurate and unimpeded flight.

The bolts of these sporting crossbows resembled short thick arrows rather than ordinary crossbow bolts. Whether they were nocked or not, I cannot say. It is possible, however, they were nocked to prevent them from slipping

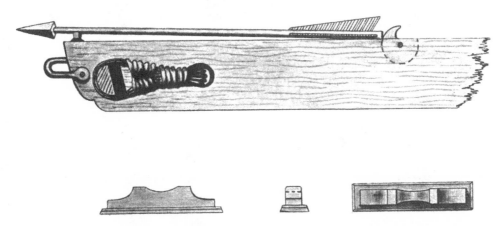

FIG. 89.—SHOWS HOW THE BOLT WAS LAID ON THE STOCK OF A CROSSBOW WHICH HAD NO GROOVE DOWN ITS CENTRE.[1]

The small ivory support which props up the head of the bolt is given full size.

off the bow-string when it was released. In the older crossbow which had a groove along its stock to influence the flight of the projectile, a nock in the butt-end of the bolt was unnecessary. As sporting crossbows were required to shoot with great accuracy, it is probable the manner of placing the bolt on the stock shown in fig. 89, was the most effective for precision.

The more ancient methods of laying the bolt of a crossbow on its stock are given in Chapter XIII.

Though the military crossbow was generally superseded in open warfare by the hand-gun about 1500, the sporting crossbow bent by a cranequin, was popular with sportsmen and foresters for over a century later. This was because of the silent discharge of the crossbow, and owing to the imperfect ignition, noise and costliness of the arquebus.

[1] The support being mortised transversely into the stock, could be moved to the right or left and then clamped by a small screw when adjusted to give a correct flight to the bolt which rested upon it.

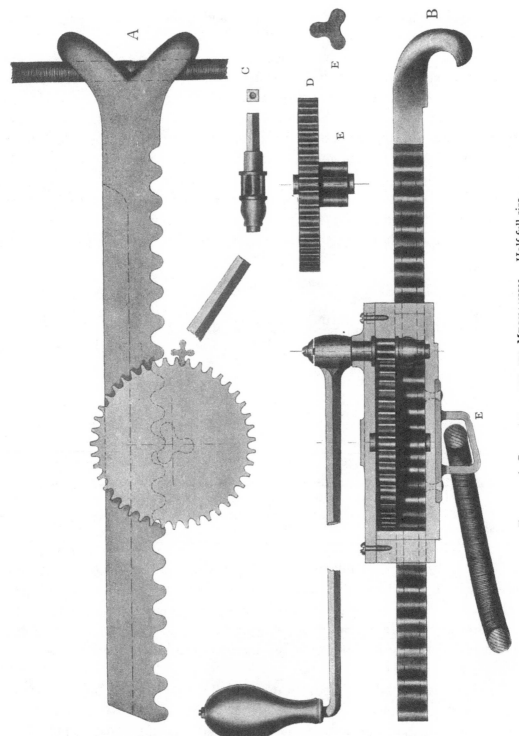

FIG. 90.—A Cranéquin and its Mechanism. Half full size.

The powerful cranequin crossbow was discarded for killing large deer, boar and wolves, about 1635. A lighter form of this weapon was, however, regularly used by hunters of game-birds, and animals of inferior size, such as chamois, roebuck, hares and rabbits till 1720–1730.

There are some beautifully constructed sporting crossbows of moderate dimensions with their cranequins, to be seen in Continental museums, which were made for the use of the chief foresters and keepers of royal domains as recently as the end of the first quarter of the eighteenth century.

CHAPTER XXXI

THE CRANEQUIN, AND HOW IT WAS APPLIED TO BEND THE STEEL BOW OF A CROSSBOW

A, fig. 90, p. 138. The surface view of the working parts of a cranequin, its casing being omitted. The claws of the ratchet bar are to be seen hooked over the bow-string.

B, fig. 90. The side view of the cranequin with its parts fitted, the casing being shown in section only.

c, fig. 90. The small spindle which is secured to and turned by the handle of the cranequin.

D, fig. 90. The side view of the large wheel.

E, fig. 90. The small wheel, with its three thick cogs. This wheel is part of the large wheel D, both wheels being made of one solid piece.

F, fig. 90. The strong metal ring that holds the thick cord loop which is placed over the stock of the crossbow when the cranequin is about to be used.

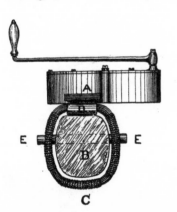

FIG. 91.—SECTION OF THE STOCK OF THE CROSSBOW, SHOWING THE POSITION OF THE CRANEQUIN WHEN IT IS FITTED FOR BENDING THE BOW.

A. The after end of the ratchet bar.
B. Section of the stock.
C. The cord loop.   D. The metal ring.
E, E. The metal pin that passes through the stock.

The under side of this metal ring rests upon the top of the crossbow when the cranequin is in position for bending the bow.

The ring is fixed below the centre of the left half of the casing, underneath the ratchet bar, D, fig. 91.

This position of the metal ring, causes the cord loop and the ratchet bar and its claws, to jointly take the great strain of the bending bow in line with the stock of the crossbow, fig. 93, p. 143.

If the ring which holds the cord loop was fixed beneath the centre of the

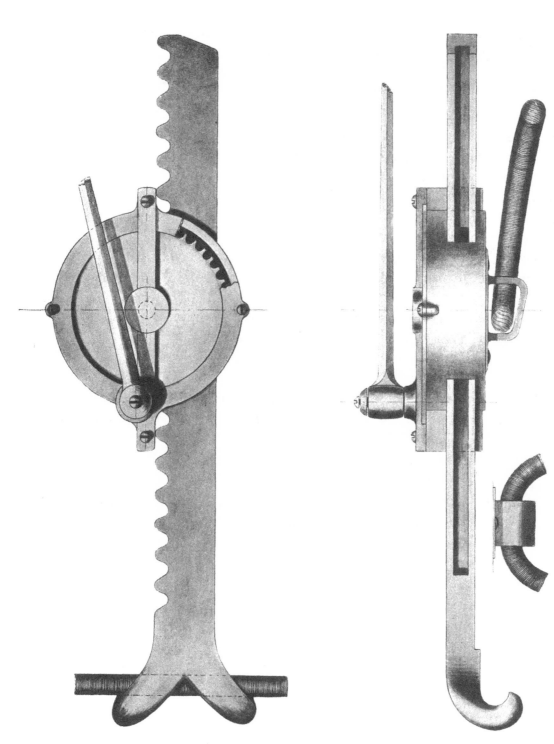

FIG. 92.—A CRANEQUIN WITH ITS MECHANISM FITTED. Half full size.

A small portion of the top of the casing is removed to show the wheel beneath it.

The end of the handle is omitted for want of space.

casing of the cranequin, the strain would be on one side of the latter, and its interior mechanism would then surely give way. The long hollow in the plain edge of the ratchet bar was cut out to reduce the weight of the instrument.

HOW TO USE THE CRANEQUIN TO BEND THE BOW OF A STEEL CROSSBOW

1. The cord loop of the cranequin is first slipped over the small end of the stock of the crossbow. It is then pushed along the stock till it comes against, and is checked from further progress by the transverse metal pin, fig. 93, opposite page.

This pin passes through the stock of the crossbow, some 6 or 7 in. behind the catch for the bow-string. It projects 1 in. on each side of the stock, and is $\frac{1}{2}$ in. in diameter.

2. The claws of the ratchet bar are next hooked over the centre of the bow-string, fig. 86, p. 133. By giving the handle of the cranequin a few turns, one way or the other, the claws can be made to move to or fro till they grip the bow-string.

3. If the handle is now turned to the right, the wheels inside the casing of the cranequin will cause the ratchet bar to slide towards the stock-end of the crossbow till the claws at the end of the bar finally pull the bow-string over the catch of the lock, fig. 93.

This operation will occupy about 35 seconds if the handle is revolved at fair speed, the force exerted by the hand in turning the handle being so slight that a strong bow can be bent by the first finger and thumb.

The action of the cranequin is very simple and powerful, though the instrument itself is so small, figs. 90, 92, pp. 138, 141.

By turning the handle, the small spindle attached to it revolves the large wheel. This large wheel causes the small wheel with three cogs, which is part of the large wheel, to work in the notches of the ratchet bar. As a result, the ratchet bar is irresistibly forced to or fro according to the direction in which the handle of the cranequin is moved.

The cranequin has no 'stop' acting on its wheels or bar, so that when the bow-string is safely hitched over the catch of the lock, a couple of reverse turns of the handle will free the claws of the cranequin from their grasp of the bow-string. The cranequin being then loose, can be quickly removed from the crossbow by pushing back its cord loop over the end of the stock. After its bolt has been laid in position the weapon is ready for discharge.

Fig. 93 shows a crossbow having its bow bent by a cranequin. The crossbow is held upright in the left hand with the fore-end of its stock upon the ground, the right hand being employed to turn the handle.

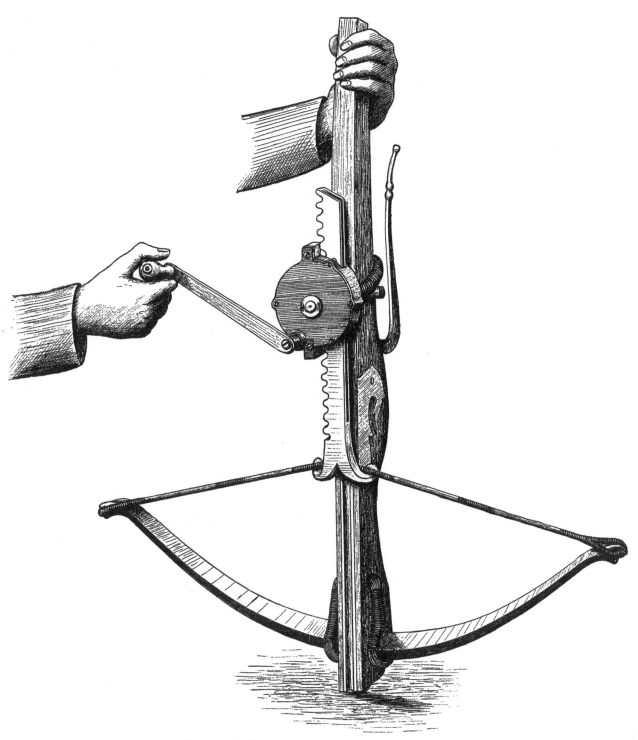

FIG. 93.—A CROSSBOW HAVING ITS BOW BENT BY A CRANEQUIN.

A cranequin crossbow seldom had a stirrup (a stirrup for the foot was not required in this weapon as it was in a windlass crossbow), but a short stock, and a bow which was usually fixed to its stock by a bridle of cord or sinew, instead of by iron clamps, fig. 87, p. 135.

For this reason, the cranequin crossbow was comparatively light and portable, even though it had a thick steel bow.

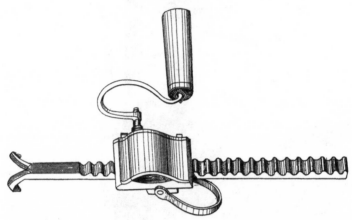

FIG. 94.—CRANEQUIN (FRENCH, END OF FIFTEENTH CENTURY), WITH METAL LOOP FOR STOCK AND COGS ON UPPER SURFACE OF RATCHET BAR.

See fig. 97, p. 151, for a soldier using a cranequin of this description.

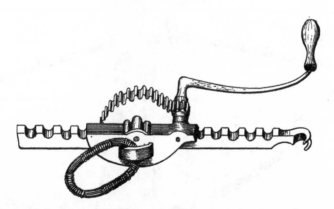

FIG. 95.—IMPROVED CRANEQUIN (GERMAN, SIXTEENTH CENTURY)

CHAPTER XXXII

THE SIXTEENTH, AND EARLY SEVENTEENTH, CENTURY SPANISH SPORTING CROSSBOW, WITH A STEEL BOW OF MODERATE STRENGTH, WHICH WAS BENT BY A CRANEQUIN

THIS crossbow was employed in Spain in the sixteenth century and till about 1635,[1] for killing deer, boar and wolves with a poisoned bolt. It was also in common use for shooting smaller animals and game birds with an ordinary bolt, and continued to be popular for this purpose till the end of the first quarter of the eighteenth century.

The powerful sporting crossbows previously described, which were carried by hunters of deer and other beasts of size, caused death to the quarry by the mere force and penetration of their heavy bolts.

In some parts of the Continent, however, particularly in Spain, a small cranequin crossbow was used for deer.

With the Spanish crossbow, it was merely necessary that its bolt should be discharged with sufficient strength to perforate the skin of a deer, the deadly poison with which the head of the bolt was smeared quickly ensuring death by mingling with the blood. The Spanish crossbow which shot a light poisoned bolt, was, therefore, of no great power. It was smaller and more convenient for the hunter to carry than the larger and stronger weapon of France, Italy and Germany, which dealt destruction to an animal by sending its bolt deep into some vital part of the body.

The use and construction of this Spanish weapon are well described by A. M. del Espinar.[2]

This author was a Spaniard, and his quaint and rare book was printed in 1644. He is the only writer who describes in some detail the crossbow used for killing deer in Spain. He explains the mechanism

---

[1] In 1644 del Espinar bitterly laments the laying aside of the crossbow in favour of the arquebus for killing deer.

[2] *Arte de Ballesteria y Monteria*, Alonzo Martinez del Espinar. Madrid, 1644.

and management of this weapon and also gives minute directions to the crossbowman concerning how he is to seek and stalk his game, whether boars, stags or wolves, under varying conditions of covert and weather.

I will now give some curious and interesting extracts translated from del Espinar, relating to crossbow-shooting as formerly practised by hunters and foresters in Spain.

———————

EXTRACTS TRANSLATED FROM A WORK ON FIELD SPORTS IN SPAIN, WRITTEN BY ALONZO MARTINEZ DEL ESPINAR, 1644

'The origin of the name Ballestero or crossbowman.

'The crossbow was much used before the introduction of the arquebus. Those who followed the chase of the larger or smaller game killed with this weapon, and except for rapid firing or when a bird was on the wing it was highly esteemed, and the use of it produced very skilful crossbowmen.

'Now the use of the crossbow has almost ceased, and with it the race of expert crossbowmen, for wings are no longer useful to birds or cunning and speed to animals, for the arquebus makes all easy for the sportsman, and thus everywhere birds and beasts are overtaken by death.

'When one or two of the many attained to the dignity of a finished sportsman, he was called a "Ballestero," that is "a Crossbowman," thus taking the name of the weapon with which he slew his quarry.

'When speaking of those who understand this art, even when they are princes, it is common to say "the King is a great Crossbowman," hence much honour follows those who practise with the crossbow.

'He who earns the name of Crossbowman must be a general sportsman, as has been already said, for it is not well that he who knows but one of the arts of hunting, whether of one kind or the other, should be so called. Therefore different names are given to other sportsmen who follow the chase, and their names accord with the kind of chase they follow.

'Some are called "Bird-catchers" because they hunt birds with snares and decoys, nets and various kinds of instruments. Others, who are called "hunters," kill large birds, rabbits and hares, with the arquebus and with wire snares.

'Others hunt partridges with a tame decoy bird, and lay snares of cords which they call "perchas." They also hunt by night with a dark lantern, which is used to drive the birds into a net. Some hunt with ferrets and nets and with pointer dogs. All these persons are known as "Hunters."

FIG. 96.—A CROSSBOWMAN WITH A STONEBOW.

Here we see a man of position with his crossbow, and his coach and horses in attendance. The birds, in this case thrushes, larks and blackbirds, are, however, being caught by means of twigs smeared with bird-lime.

From a ' Natural History of Birds,' by G. Pietro Olina. 1622

'There are some who are called "Foresters," who pursue larger game and surround and kill it with the arquebus and with dogs.

'Only those sportsmen are called "Ballesteros" (Crossbowmen) who hunt every kind of game. The Ballesteros hunt the stag and deer on horseback, they know how to stalk and they know the tracks and habits of all wild animals and where they may be killed. The Ballesteros make hunting parties for every kind of animal and they know the haunts and habits of each one, according to its nature, and everything that belongs to the craft of forestry and hunting.'

OF THE CROSSBOW AS AN INSTRUMENT OF THE CHASE

'The crossbow is safer in its management for the life of man than the arquebus, for a fatal accident has never happened through the breaking of the bow or of the cord of the crossbow, which are the two possible dangers and which fail sometimes, when they may inflict injury but nothing serious.

'The crossbow is in many ways superior to the arquebus. It is more silent, and in a herd it kills but does not alarm. It gives a dumb blow if he who uses it is dexterous. It is not so with the arquebus, which by its report alarms and scatters the herd. The crossbow is cleaner and less costly in its use. It is more effective than the arquebus and when once prepared for discharge never fails. Failure, on the contrary, too often happens with the arquebus. The crossbow also kills the greater as well as the lesser game. Anciently this instrument was more used in Spain than in all the rest of the world, therefore the best master-makers of crossbows were found in Spain rather than in other kingdoms.

'We will now give the names of the appendages or ornaments of the crossbow, and of the iron and horn parts of which it is composed.

'In order that the connoisseur may know for the future the marks of the best makers, those crossbows which are the best and most ancient are marked with a cross.

'The elder Azcoitia made the "tablero" or wooden stock and also the "gafa" used for bending the bow,[1] and he put his name on the trigger of the stock and on the "gafa."

'Pedro de la Fuente made both stock and "gafa" and put his name where Azcoitia put his.

----

[1] This was the 'cranequin' or ratchet winder, Chapter XXXI., which in the Spanish crossbow was sometimes permanently fixed to the stock.

CROSSBOWMEN WITH CRANEQUIN CROSSBOWS.

From paintings by Holbein and other Medieval Artists.

'Cristoval de Azcoitia, grandson of Azcoitia the elder, made stock and gafa and put his name where others had done, calling himself Azcoitia the fourth.

'Juan Hernandez made stock and gafa and put his name also on both parts. Juan Peres de Villadiego did the same.

'Juan de Azcoitia only made the stock.

'Uzedo made stock and gafa; Juan Criado only the stock; Hortega, stock and gafa.

'Of all the great master-makers of crossbows, none remain save Juan de Lastra, who serves his Majesty [Philip IV] in the office of crossbow-maker. There are many other makers but these are the ones who have achieved most fame.'

'The masters who have made strings for crossbows are, first and best, Louis Moreno, then Juan Blanco. The elder Puebla Alanis, Grajeras the deaf man of Zamora, Munoz of Getafe and others have made them in this kingdom and in Biscay.'

THE IRON AND HORN OF WHICH THE CROSSBOW IS MADE AND ITS APPOINTMENTS

'The crossbow has a stock. The irons which furnish the sides of the stock —at the part where the nut is—are called the cheeks of the crossbow.[1] These irons are sunk into the wood and adjusted so as to be level with it.

'Two irons which surround the centre of the bow near the head of the stock are called the "flowers," one of these irons is on each side of the stock.[2]

'The trigger which frees the nut that holds the string of the crossbow, is the long iron underneath the stock.[3]

'The nut which holds the crossbow string when it is stretched, is made out of the horn which stags have on their heads at the bottom of their antlers, and there is nothing so strong for this purpose in any other animal.[4]

'This nut of horn has also a steel catch which meets the point of the trigger inside the stock, and these fit one with another when the crossbow is bent.[5]

'The hollow inside the stock in which this nut revolves, is called the "box,"[6] and this hollow has a horn fitting round it which is called the breast piece.

'On the top of the stock, forward of the nut, there is a long grooved horn. In this groove the arrow is laid ready for discharging after the bow is bent.[7] This horn is called the "canal," and the part of the stock behind the canal and the nut is called the tiller or handle of the crossbow.

----

[1] The lock-plates, fig. 56, p. 99.   [2] The bow-irons, fig. 59, p. 103.   [3] The trigger, fig. 55, p. 98.
[4] The crown.   [5] Fig. 52, p. 97.   [6] The socket, fig. 50, p. 96.   [7] The groove, fig. 64, p. 108.

'A little iron ring at the head of the stock is called the "etrivo." These are the horn and iron parts of this weapon, excepting the steel bow and the "gafa" for bending it.[1]

'To complete the crossbow it should possess the following qualities :

'It should be safe for the face of him who shoots it, so that it may not hurt him. It should be easy to discharge and sure not to go off before its time. The arrows should fly straight, for in this lies the great excellence of a crossbow and its certainty and good aim. When the arrow does not go straight but flies aside or in a sinuous way, it is not likely to reach its mark, but when it flies straight all goes well.

'We will now mention the causes which prevent the arrow from flying straight.

'When the bow is not well placed in the stock. If the arms of the bow on either side of the stock are not level, but one arm is longer on one side of the stock than it is on the other side, then the force of the bow is not equal, because the arm which is longer overcomes the other arm and the arrow cannot go straight.[2] This fault in a crossbow is called "alti-bájo."[3] It may be remedied by replacing the bow evenly in the stock, for if the arms of the bow are not of the same length, by the thickness of a thread on each side of the stock, it will fare ill with the arrow with which the crossbow is loaded.

'Some crossbows are unruly and injure him who shoots with them. This is caused in two ways. The principal one is, that the bow has too much steel and the stock too little wood, so the excessive strength of the steel overbears the wood and causes the stock to recoil upon the face, whereby there is wounding and offence. It should be seen, therefore, that the stock is of the same weight as the steel bow. The bow should not be unruly or too strong in its discharge but rather working quietly than striving with all its might, for this latter causes recoil or kick. If there be no inequality in these things, the crossbow is safe and the man is happy when he shoots with it.

'When all these matters are adjusted, the crossbow may still be unruly from the bow being loose in the stock, which causes much recoil when the arrow is shot off. This is called "tener dientes" or to have teeth.'[4]

'There are also two other reasons why a crossbow may shoot its arrows badly. The first is, that the string of the bow is placed too tight against the

---

[1] The Construction of the Crossbow, pp. 92 to 128, will elucidate del Espinar's description of the parts of the Spanish weapon of chase. The 'Etrivo' or small iron ring at the fore-end of the cranequin crossbow, was attached to the stock as a means of suspending the crossbow to a hook in a wall or, in the case of mounted crossbowmen, to a metal loop fixed in the saddle.

[2] The author here gives a lengthy description of how to measure with a thread the centre of the bow as applied to the centre of the stock.

[3] 'Ups and downs.'

[4] Anglice, 'To set the teeth on edge.'

FIG. 97.—SPANISH CROSSBOWMEN, WITH CRANEQUINS AND GOAT'S-FOOT LEVERS.

From an album containing specimens of Spanish soldiery from the earliest times to the present.
By Count de Clonard.   Published by order of the Spanish War Office.   Madrid, 1861.

surface of the stock, wherefore it presses so hard upon the stock that the string cannot play freely along the groove in which the arrow is laid.   The centre of the string does not then strike the arrow in the middle of its butt, but rather below it, so the arrow goes away with a sinuous flight.   In such a case a cross-bow is said to be " loaded " or " weighted."

'The second evil is, when the string of the bow is raised just clear of the stock, because then the centre of the string strikes the arrow rather higher than the middle of its butt and so, instead of shooting it off properly, it drives it downward.

'The arrow from a crossbow will also fly ill if it press against the stock as the string sends it off, for if the arrow is to fly well it must only rest at its head and at its butt-end upon the groove of the stock.

'There are many different sights in crossbows, because men generally take aim differently and so order the sights to be made in various ways to suit their individual tastes.   The most perfect crossbow is the one which has the stock straight from the head to the end of the handle.

'In order to take aim with the crossbow, the shooter must grasp the handle of the stock with his right hand and place his thumb over the upper surface of the stock, then, as he holds the stock and the trigger in this hand, he should raise his thumb close to his eye.[1]   When the head of the arrow can be seen above the top of the thumb, he takes aim as he chooses and in this way he will strike his game ; but the thickness of a " real of eight "[2] will give a shot a finger length higher or lower than it should be.

'It is needful to know that the direct flight of the arrow of the crossbow is generally twenty-five paces,[3] up to which distance it hits very surely, but after five paces more it will strike lower according to the strength of the bow.   The weaker crossbows will strike two finger lengths lower at thirty paces than they do at twenty-five paces, and the stronger one finger length lower.

'The shooter must aim according to his knowledge of the strength of his crossbow and the distance of the game at which he shoots, but this is not a matter which can be settled here.

'The art of using the crossbow is so greatly lost here in Spain where formerly such beautiful things were done with it, that I have desired to name its parts, in order that these should be preserved in memory for the benefit of lovers of curiosities, and because the crossbow is the best instrument with which to teach princes from their childhood.   Also the crossbow teaches all the

---

[1] The author does not mention the left hand, which of course grasps the stock forward and near and below the bow.

[2] A silver piece of about the thickness of a dollar and an ounce in weight, comprising eight small coins in value.   [3] Point-blank range the author means.

delicacies of taking aim in readiness for the time when princes shall use the arquebus and shoot with ball.

'Royal grandeur may always occupy its leisure in slaying wild beasts, and if the King is skilful in the use of the crossbow he will easily learn to master fire-arms.'

FIG. 98.—A CROSSBOWMAN WITH A STONEBOW.

Though in this view the hunter may be seen with his crossbow, the birds are being chiefly taken by limed twigs fixed round the owl which attracts them.

From a ' Natural History of Birds,' by G. Piétro Olina. 1622.

OF THE ARROWS WHICH ARE SHOT FROM THE CROSSBOW, TO KILL STAGS, ROE-DEER,
WILD BOARS AND SMALLER ANIMALS

'The arrows of the finest kind are called "Jaras," because they are made of Jara[1] wood.   They will strike at 150 paces or more.

[1] Cistus, or rock rose.

'These are anointed with a poison called "The Crossbowman's herb." This kind of arrow is smeared with the juice of this plant from the neck of its iron point for five or six finger-breadths down. Then the arrow is covered with a little strip of very thin linen, which wraps the shaft round and round and adheres to it over the poison without the need of any tying. The head of this arrow is of steel, square and pointed, and the neck very thin.

'There are also arrows called "Sostrones" for night use. These are large and heavy, so that they cannot be shot far from the crossbow and are therefore easy to find when they are discharged at rabbits and hares by moonlight. They are also used with a dark lantern to kill pigeons on their roosts in the trees at night.

'Then there are other arrows for killing partridges, which are a hand's breadth longer and have an iron knob at the head.[1]

'Other shafts are called "Pasadores," and are thicker than the ordinary arrows. Some shafts are known as "Rallones," the points of which are like a chisel. There are also some termed "Saetones," used for shooting at leverets or young rabbits; they are longer than the ordinary arrow and very sharp, and in the middle of the shaft is a small bar, so that when a rabbit is struck it cannot go down its burrow.

'The best masters who have made arrows for crossbows in Spain are :

Christoval de Escobar, who served the Lord Kings Philip II. and III.
Juan de Escobar, his son, who also served the Lord King Philip III.
Juan Martinez, Juliers Perez, the two Renedos and Acacio.'

HOW TO MAKE THE POISON FOR THE ARROW OF THE CROSSBOW

'This decoction is made of the roots of the white Hellebore, which should be gathered towards the end of August as it is then at its best season and strength. The smallest roots and those which are darkest in colour and turning yellow are the best from which to make the poison, the whiter roots not being so strong. These roots may be gathered in the mountains of Guadarrama and in those of Bejar. They are like small turnips and the thinnest and most hairy are the best.

'The way to treat them is to take off all earth and any kind of viscous matter which may adhere to them and then to wash them well. After this they should be pounded and placed under a press to extract all their juice,

---

[1] The round blunt head was used to prevent the birds being damaged, as would be the case if they were perforated by a sharp arrow.

which will have to be carefully strained and then put over a fire to boil. All froth and viscosity which may remain must be skimmed off the juice. When this is done, the juice must be strained again and then set in the sun from 10 o'clock in the morning till the day declines.

'This process will have to be repeated for three or four days or more. Each day before the juice is set in the sun it must be strained, when it should be like syrup, and of the same colour but thicker. If you put a straw or a bit of stick in it, it should adhere to it, and that which gathers together most quickly and which if smelt makes people sneeze violently, is the strongest.

'Some people who make this decoction boil it, instead of exposing it to the sun, but this decoction is not so strong as that which is set in the sun.

'The poison may be tried on a chicken or a young pigeon to see if it is right. Take a needle with thread, wet the thread in the mixture and pass it through the sole of the chicken's foot between the skin and the flesh till it bleeds, then, in the time of saying "Credo," the bird will nod and in a very short time will die. The same will happen with a cat or any other animal if the decoction is good.

'I have seen such a thing happen, that when a stag receives a wound from a poisoned arrow it runs a hundred paces, more or less, and then turns its head towards the place whence it received the shot; in a very short time, during which the animal stands still if it can do so, the poison reaches its heart and so puts an end to its life.

'This may be known by the animal beginning to cough and toss its head and vomit; then in an instant it is dead.

'Let it not be thought that without the poison the animal dies of the wound of the arrow though it has received it in a mortal part.

'If the animal be wounded only in the hoof, and if it bleeds at all, this poison will work into the blood and death will soon come. There are other decoctions made from other herbs which are slower in their effects though they also kill. There are also others which make an animal stagger, reel and vomit only, and the more they vomit the sooner they recover.

'The places where the poison is slowest to take effect, are wounds in the stomach, for as there is no blood in the stomach the strength of the poison is spent in its contents.

'The animals which most quickly die from the poison are those of most choleric temper. The wild boar, the wolf and the cat, die very quickly.'

FIG. 99.—SHOOTING BIRDS AT NIGHT WITH A STONEBOW.

The pigeons or other birds were dazzled with a lantern and then killed whilst at roost.
If they flew out of the trees, they were knocked down with sticks which had heads like racquets.

From a 'Natural History of Birds,' by G. Piétro Olina. 1622.

CHAPTER XXXIII

THE SIXTEENTH-CENTURY STONEBOW, WITH A LIGHT STEEL BOW, WHICH WAS BENT BY MANUAL POWER ONLY

ARBALETE A JALET—STONEBOW—PRODD—LATCH

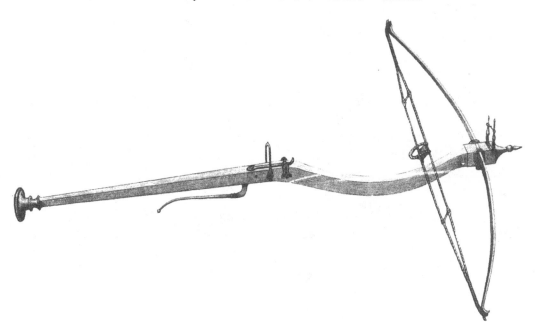

FIG. 100.—A SIXTEENTH-CENTURY STONEBOW WHICH WAS BENT BY HAND.

THIS was a sporting crossbow with a steel bow which was fitted with a double string. In the centre of the double string, a little pocket of interlaced twine—called the cradle—was fixed to hold the pebble the crossbow discharged.[1]

With our ancestors, the stonebow, to some extent, took the place of the small rifle we now use for rooks and rabbits.

[1] The stonebow was used in sport and never in warfare. Many of the stonebows which were made for the nobles and princes of the sixteenth century were splendidly decorated, the metalwork, engraving, carving and inlaying to be seen on their stocks being often of lavish beauty. To such an extent was this embellishment carried, that the stonebow was sometimes even fitted with a bowstring formed of chased steel links, instead of the usual cord.

The original stonebow appeared about 1500 and soon became very popular for killing game birds on the ground by day or, by the aid of a lantern, pigeons roosting on the trees at night.

Stonebows were made of various dimensions; some, intended for ladies, were so light that they could be held and aimed with one hand, while others had bows of two feet and more in length. These latter were put to the shoulder and were of sufficient strength to knock over a rabbit, or a pheasant or pigeon sitting on the bough of a tree.

The manipulation of the stonebow is well described in the poem 'Le Plaisir des Champs,' by Claude Gauchet d'Ampmartinois, printed at Paris in 1583.

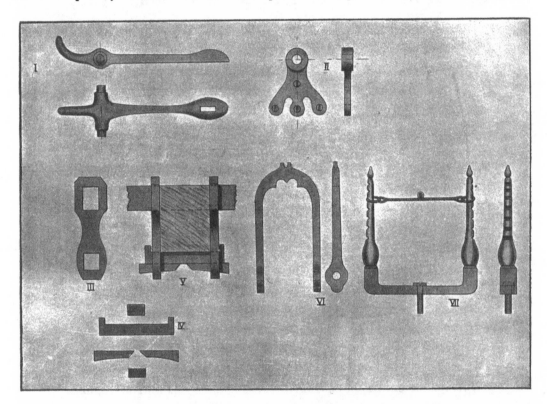

FIG. 101.—THE PARTS OF THE LOCK OF THE PRIMITIVE STONEBOW, ITS SIGHTS AND THE MANNER OF FIXING THE BOW TO THE STOCK.

I. The catch for the bowstring. II. One of the two uprights in which the spindle of the catch hinges. III. One of the side straps that secure the bow to the stock. IV. The wedges that secure the straps and the bow to the stock. V. The bow secured by the straps and wedges to the stock. VI. The back sight. VII. The fore sight. (For these parts fitted, see fig. 100, previous page, and fig. 102, opposite page.)

A literal translation of the passage runs:

'Then with crossbow in hand I draw near, and placing a ball in its sling[1] and the loop[2] upon the nut[3] of the lock, I bend the bow. Through the little

---

[1] The pocket that holds the pebble or bullet.   [2] The loop attached to the pocket.

[3] The catch for the bowstring.

sight-hole [1] I aim at the bird and cover her with the bead.   Then standing steady I press the key.[2]   The steel bow recoils with great force and drives the ball towards the bird, which falls dead to the ground.'

The oldest examples of the stonebow, such as those of the sixteenth, and first quarter of the seventeenth century, (fig. 100, p. 157,) had their bows bent by manual power without the assistance of a lever.

The steel bow of a primitive stonebow of this description, was, therefore, comparatively weak, else it could not have been bent by the hand alone.   The

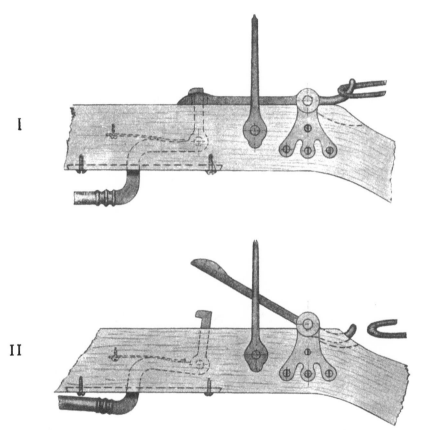

FIG. 102.—THE PARTS OF THE LOCK OF THE PRIMITIVE STONEBOW, AS FITTED INTO THE STOCK.

I. The bowstring hitched over the catch of the lock. The catch is secured by the notched top of the trigger passing through the opening in its rounded end. II. The bowstring escaping from the hook of the catch. The trigger having been pressed upwards, its notched top moves forward and thus allows the catch to escape from its grip.

early stonebow had the fore-part of its stock curved downwards, (fig. 100, p. 157,) with a view to allowing the bow-string to recoil without being checked by friction against any part of the stock.

In these stonebows of the sixteenth century, which may always be recognised

___

[1] Peep-sight.   [2] The long trigger used at the time.

by their curved stocks, the more modern plan of canting up the ends of the bow from its centre, so as to lift the bow-string clear of the stock, does not seem to have been thought of.

The method by which the pebble was enabled to pass forward clear of the centre of the bow, was by fixing the latter with an upward slant in the stock.

CHAPTER XXXIV

THE SEVENTEENTH-CENTURY STONEBOW, WITH A THICKER STEEL BOW, WHICH WAS BENT BY A LEVER FIXED IN ITS STOCK

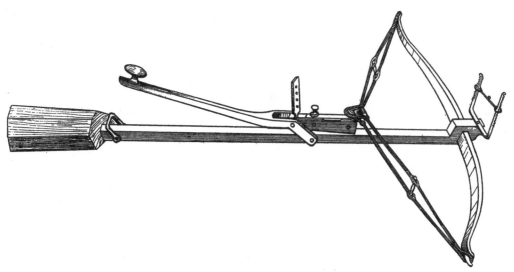

FIG. 103.—STONEBOW WITH A LEVER FIXED IN ITS STOCK.
Seventeenth century.

IN the first half of the seventeenth century, the stonebow was fitted with a lever to bend its bow and stretch its bow-string. This was a most useful addition and one that allowed of a considerably more powerful bow than had previously been possible.

A stonebow of this kind was, of course, capable of throwing a pebble with much more force than was attainable from a bow, as described in the last chapter, which was bent by the unaided hand.

The general mechanism of the improved stonebow of the end of the seventeenth century (fig. 104, next page), was the same as that of its successor the modern bullet crossbow of the nineteenth century, Chapter XXXVII. Even its sights, lock, lever, bow-string and other parts were

similar in their action to those of the modern weapon, with the exception
that its bowstring was set free by pressing a button instead of by pulling a
trigger.

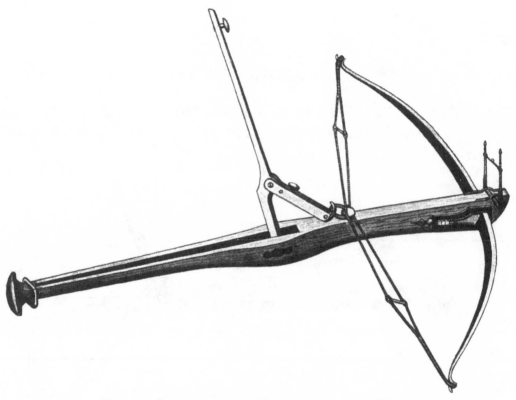

FIG. 104.—STONEBOW WITH A LEVER FIXED IN ITS STOCK.

End of seventeenth century.

The lever is here shown hitched to the bow-string. The lever, when pulled down, was secured by
sliding the loose ring at the butt end of the stock over its end.

The crossbow being then bent, was discharged by pressing the knob on the top of the lock.

CHAPTER XXXV

*THE SEVENTEENTH AND EIGHTEENTH CENTURY SPORTING AND
TARGET CROSSBOW, WITH A LIGHT STEEL BOW, WHICH WAS
BENT BY A WOODEN LEVER*

THIS handsome little crossbow (fig. 105, p. 165), was used in the latter half
of the seventeenth and in the first half of the eighteenth century.

After the powerful crossbow for killing deer had been supplanted by
the arquebus, the makers of crossbows turned their attention to the production
of weapons that served for target-practice and for use on game and other
birds, as well as on small animals such as hares and rabbits.

These light crossbows were of excellent design and construction and often
beautifully decorated; they discharged their bolts with great accuracy up to fifty
yards and with considerable force.

Though their steel bows were small in comparison with those of the
bolt-shooting crossbows previously described, they were stout for their length, and
capable of sending their miniature quarrels to a distance of 200 yards.

This weapon was very popular on the Continent for killing game birds and
small animals till about 1720, and for target-shooting till a much later date.

In mediæval target-shooting, the bolts of crossbows were commonly aimed
at a disk of white paper or other conspicuous object,[1] placed, at 50 yards distance,
against the centre of a tightly rammed butt of grass sods, about 5 ft. square on
its face.

The bolt of a crossbow, being so short in comparison with the arrow
of a longbow, was very liable to be lost if discharged so as to fall on level
ground, by reason of its penetrating beneath the surface of the soil.

[1] The half of an oyster shell, its white side, of course, outwards, was a favourite mark to set up against
a butt for crossbow-shooters to aim at.

The upright butt of earth prevented the bolt being lost in this way.   The usual plan in target-shooting with the crossbow, was to have two butts connected by a path excavated to a depth of about 2 ft.

The crossbowman passed to and fro along this path, as he shot his bolts first at one butt and then at the other.

On each side of the path, the ground was smoothly sloped and turfed to accommodate competitors and spectators.

The reason of this path, was to enable the crossbowman to hold his weapon in a level position as he took aim.   To achieve this attitude, the sunk pathway was adopted, as without it a very high and large butt would have been necessary to bring the eye of the marksman and the centre of the butt in a line parallel with the ground.

Fig. 14, p. 32, shows crossbow-shooting at the butt, together with a covered gallery for service in wet weather.   The crossbowmen depicted are using cranequin crossbows.

DIMENSIONS OF THE SMALL SPORTING CROSSBOW, FIG. 105, OPPOSITE PAGE

The steel bow is 2 ft. long.   At its centre, it is $1\frac{1}{8}$ in. wide and $\frac{3}{8}$ in. thick.

The stock is 2 ft. 3 in. long, and $1\frac{1}{4}$ in. wide across its top surface where the bolt is laid.

From the fore-end of the stock to the back of the centre of the bow, the space is $2\frac{5}{8}$ in.

From the inside of the centre of the bow to the catch for the bow-string, 8 in.

The draw of the bow-string, from a state of rest to the catch of the lock, 5 in.

From the extremity of the fore-end of the stock to the catch which secures the stretched bow-string, 11 in.

THE WOODEN LEVER

As the steel bow of a serviceable target—or light sporting crossbow could not be bent by hand alone, a clever and simple contrivance—an adaptation of the goat's-foot lever previously described, was employed to do this, fig. 105, opposite page.

This apparatus was small and light, weighing only $1\frac{1}{2}$ lb., and was able to bend easily and quickly a steel bow of moderate size ; one, for instance, that was not so strong as to require a windlass or a cranequin to stretch its bow-string.

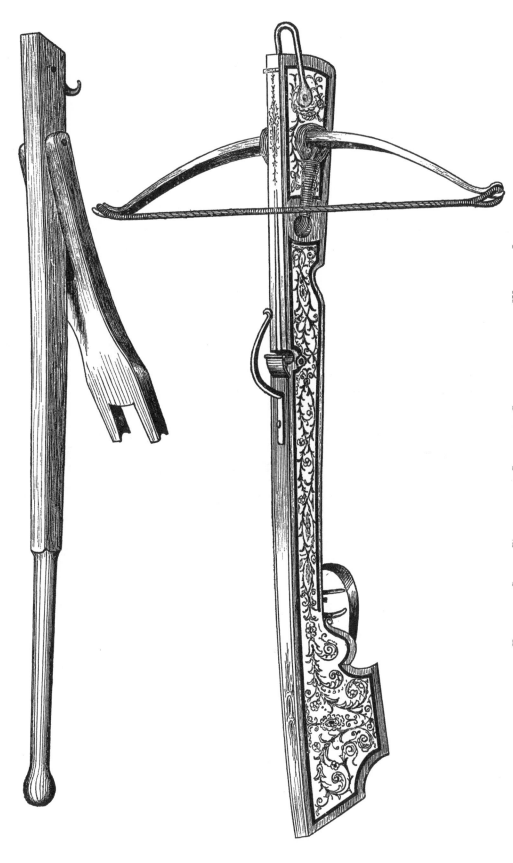

FIG. 105.—SMALL TARGET AND SPORTING CROSSBOW AND ITS WOODEN LEVER.
German, end of seventeenth century.

The length of the handle of the lever is 2 ft. It tapers from a diameter of 1 in. at its rounded end, to a width of 1¼ in. and a depth of 1½ in. at its fore-end.

The wide hinged piece pivoted to the handle, is 9 in. long to its swivel-pin, its total length being 10 in.

This piece is 2½ in. wide and 1 in. thick. It is slightly curved, fig. 106, opposite page.

The ¼ in. swivel pin by which the hinged piece is attached to the handle of the lever, is 6 in. from the fore-end of the latter.

The metal hook to be seen in the fore-end of the handle, swings loosely in a small cavity. The ¼ in. pin for this hook is 4½ in. from the pin on which the flat piece is hinged.

This form of lever for stretching the strings of small crossbows with steel bows, was no doubt suggested by the ancient goat's-foot lever. It is, however, in some respects a more convenient device than the goat's-foot lever, as it pushes the bow-string of the crossbow to the catch of the lock instead of pulling it there, and also works without friction.

Levers on this principle, made of metal, are used with Continental target-crossbows at the present day. See fig. 156, p. 217.

HOW TO USE THE LEVER TO BEND THE CROSSBOW, FIG. 106, OPPOSITE PAGE

The metal hook of the handle is slipped into the small iron loop fixed at the fore-end of the crossbow. The rounded notches in the short props of the swinging part of the lever are then rested against the centre of the bow-string, fig. 106.

The left hand grasps the crossbow near its fore-end, and the butt of the stock is placed upright upon the ground.

It will be easily understood that by now pressing the handle of the lever downwards with the right hand (fig. 106), the bow-string will be forcibly pushed along the stock, till it reaches and is gripped by the catch of the lock.

The lever being then removed, by unhooking it from the fore-end of the crossbow, the weapon is ready for use.

The bolt is arranged on the surface of the stock as shown in fig. 89, p. 137,

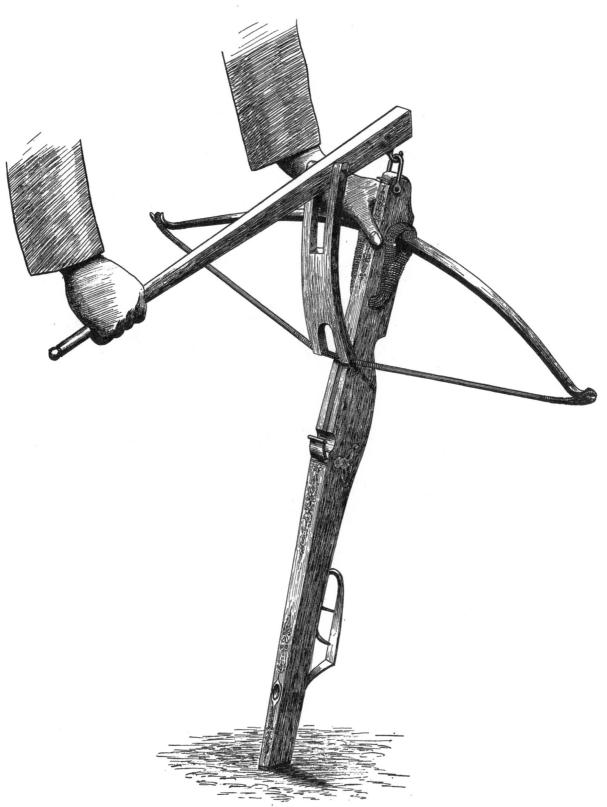

FIG. 106.—A SMALL SPORTING CROSSBOW BEING STRUNG BY ITS WOODEN LEVER.

except that in this crossbow there is a slight groove near the catch of the lock, to assist in keeping it in position.

The groove is here necessary, as this form of catch for holding the bow-string has no claws (as had the old revolving nut) between which the butt of the bolt can be placed to secure it from falling sideways off the crossbow.

The bolt was held from slipping forward (when the crossbow was directed downwards) by the piece of curved horn shown in fig. 105, p. 165.

This piece of horn acted as a spring that pressed lightly on the butt-end of the bolt, when the latter was laid on the stock after the bow was bent. See also fig. 109, p. 170, and notes thereon.[1]

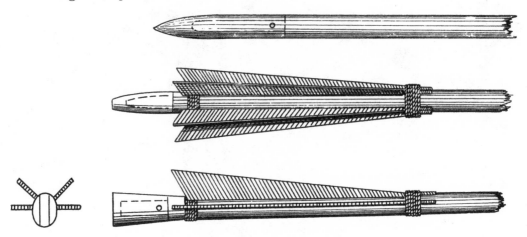

FIG. 107.—THE BOLT WITH FOUR FEATHERS USED WITH THIS CROSSBOW.

Length 12 in.; diameter of shaft ⅜ in.; height of butt-end ½ in. The butt is sheathed with brass. The head of the bolt is made of steel.

[1] When aiming his weapon, the crossbowman grasped its butt-end with the fingers of his right hand ; the first finger pulling the trigger. He placed his thumb in the small oval recess to be seen on the surface of the stock (fig. 106), to assist him to hold his crossbow firmly, and in a level position.

CHAPTER XXXVI

THE SIXTEENTH-CENTURY IMPROVED LOCKS, WHICH WERE FITTED TO SPORTING AND TARGET CROSSBOWS THAT DISCHARGED BOLTS

IN the small crossbow shown in fig. 105, p. 165, the bow-string was hitched in a sloping notch cut across the surface of the stock. The notch was protected with ivory to save it from damage by the friction of the bow-string.

When the bow was bent and the bow-string was in position in the notch, the broad flat top of a swinging catch snapped down and prevented the string from escaping. A lever inside the stock interlocked with the lower end of the swinging catch, and in this way its flat top was held fast over the bow-string.

A small independent safety lock with a trigger of its own, acted upon the lever which secured the catch. The crossbow could not be shot off till this small lock was cocked, this being done only just before aim was taken, figs. 108–113, pp. 170–173.

The great advantage of this kind of crossbow lock was its safety from accidental discharge and the instantaneous loose that a slight pull of its trigger gave to the bow-string. This free and easy release of the bow-string was, of course, of much assistance in aiming correctly, whether at the target or at game.

In the crossbow that was used previous to the sixteenth century the long trigger with which it was fitted, being more of a lever than a trigger, required some pressure of the hand to force its point out of the notch of the revolving nut to set free the bow-string, fig. 55, p. 98.

THE MECHANISM OF THE IMPROVED LOCKS OF SIXTEENTH CENTURY
CROSSBOWS

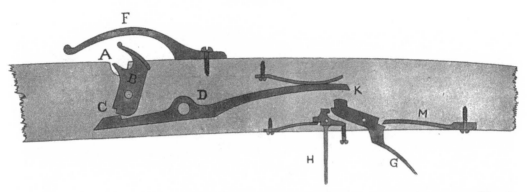

FIG. 108.—THE LOCK AS IT APPEARS BEFORE THE BOW-STRING IS STRETCHED TO THE NOTCH.
Half full size.

Fig. 108. As the bow-string is forced by its lever along the stock, it finally slips over the slope of the notch at A.

As the string drops into the notch, it presses the projection at A, of the swinging catch B, downwards. This causes the stepped end of the catch B, and the stepped end of the lever D, to interlock at C, as seen below in fig. 109. At the same moment the broad flat top of B, falls over the bow-string E, and holds it from escaping upwards out of the notch, figs. 109, 110.

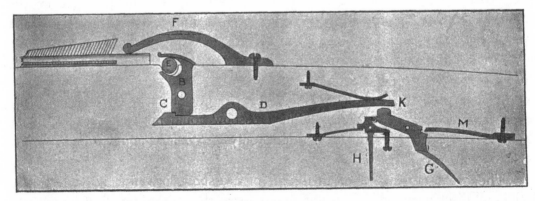

FIG. 109.—THE BOW-STRING E, SECURELY HELD IN THE NOTCH IN THE STOCK BY
THE TOP OF THE CATCH B. Half full size.

Fig. 109. The bolt is now placed on the stock; its butt-end, it will be seen, not quite reaching the notch.

The rounded end of the piece of curved horn F, (⅜ in. wide,) presses lightly on the butt of the bolt and prevents it from falling off the crossbow should the latter be aimed downwards. The top of this piece of horn has a V-shaped nick cut along its centre at F, (see dotted line,) to act as a back-sight. The point of the bolt acts as a fore-sight. As the bow-string is forced to the notch by the lever of the crossbow, it is pushed under the end of the horn F. Though the bolt is now on the stock, the bow bent and the string stretched (fig. 109), the crossbow is safe and cannot be discharged without further manipulation.

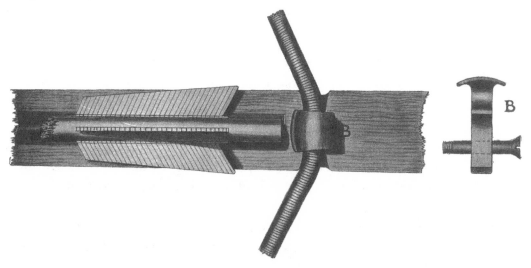

FIG. 110.—SURFACE VIEW OF THE STOCK, WITH THE BOWSTRING SECURED BY THE CATCH OF THE LOCK AND THE BOLT IN POSITION. Not to scale.

Fig. 110 shows the top of the catch B. The horn F, is here omitted to avoid confusion of details. The front view of the catch B, and its swivel-pin, is given separately.

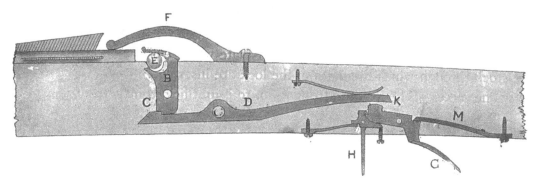

FIG. 111.—THE LOCK COCKED. Half full size.

Fig. 111. To prepare the crossbow for discharging, pull back the cocking lever (G, fig. 109), till its stepped end snaps into the notch at the top of H. G and H will then be interlocked, as shown in fig. 111.

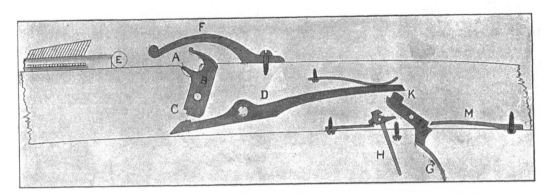

FIG. 112.—THE EFFECT OF PULLING THE TRIGGER H, TO DISCHARGE THE CROSSBOW.
Half full size.

Fig. 112.   By pulling back the trigger H, the cocking lever G, is set free, and is at once jerked forcibly upwards by the strong spring M.   The result of this is, that the projection on the top end of G, strikes a smart blow at K, underneath the long end of the lever D.

This impact of G below the long end of D, causes its other end at C, to drop down and thus instantly to disengage with B, where the two pieces were previously interlocked at C, as in fig. 111.

The catch B, being now free to swing has no further hold at its top on the bow-string E, as it had in figs. 109, 110, 111.

The bow-string E, having nothing to detain it flies out of the sloping notch and propels the bolt along the stock, fig. 112.   The lock of the crossbow then returns to the position given in fig. 108.

By twisting in or out the small screw to be seen between G and H, the trigger H, can be regulated to any pull, however light.

The pieces G, H, with their pins and springs, work in a metal casing attached to the inner face of the trigger-plate which closes the opening under the stock through which the parts of the lock are inserted.

The pieces of the lock are from $\frac{5}{16}$ in. to $\frac{3}{8}$ in. thick, transversely.

―――――――――

As the steel bow of a large sporting crossbow that was bent by a cranequin or a windlass, was far more powerful than the bow of a crossbow used for target practice and for killing birds and small animals, the former required a stronger arrangement for holding its bow-string than the notch and catch above described.

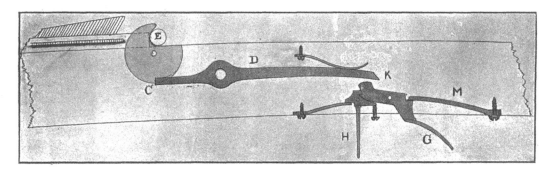

FIG. 113.—THE LOCK OF THE SIXTEENTH CENTURY SPORTING CROSSBOW OF LARGE SIZE.
Half full size.

In fig. 113, we have the revolving ivory or steel nut common to all mediæval crossbows that had powerful steel bows. We also have the usual lever D, with its point fitting into the notch of the nut at C. In this case, however, the lever D, is cut off short inside the stock, instead of being prolonged into the old-fashioned outside trigger which was pressed by hand to free the bow-string, as in fig. 55, p. 98.

The rest of the mechanism of this lock is identical with that given in figs. 108–112, and acts as follows.

On the trigger H, being pulled G is set free, and the projection at the upper end of G, deals a sharp blow at K, underneath the long end of D, fig. 113.

This blow of G below D, at K, at once causes the other end of D to drop out of the notch at C, in the circular nut.

The nut being then free to revolve, the stretched bow-string E, instantly leaves its claws as these turn over, and propels the bolt.

In this lock we have the same advantages of safety and the same easy loose of the bow-string as in the one previously described, the principle of both locks being very similar.

PART III.

THE CONSTRUCTION AND MANAGEMENT

OF

CROSSBOWS (*Continued*)

MODERN

WITH AN ACCOUNT OF CROSSBOW-SHOOTING AT THE TARGET OR BIRD,
AS NOW AND FORMERLY PRACTISED ON THE CONTINENT

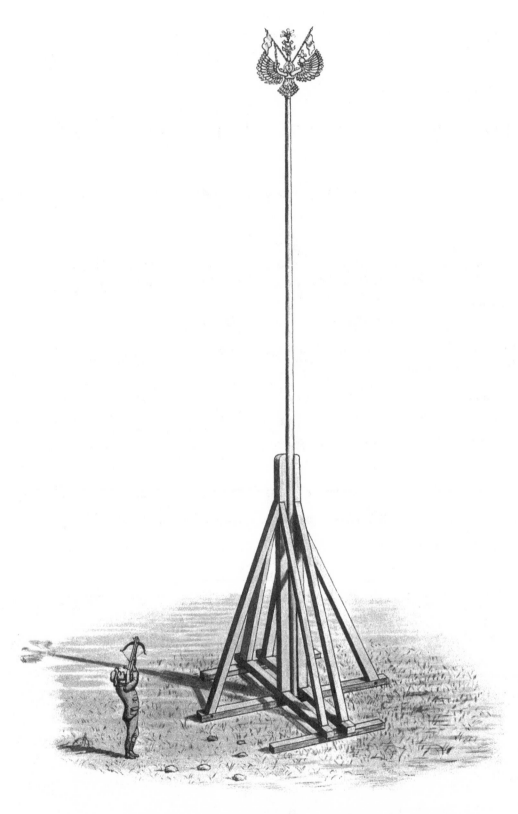

Fig. 114.—Crossbow-shooting at the Bird, as now practised in Saxony.
See Chapter XLVIII.

CHAPTER XXXVII

THE BULLET-SHOOTING CROSSBOW—ENGLISH

THIS bullet crossbow may be termed modern in comparison with the crossbows hitherto described, though the period of its popularity dates so far back as the years 1800–1840. As before stated, the bullet crossbow is a reproduction of the stonebow of the sixteenth century (fig. 100, p. 157), which it closely resembles, save that it has a considerably more powerful bow, a lock of better design, and a lever attached to its stock for bending its stronger bow.

The bullet crossbow, a handsome and effective weapon of 6 lb. to 7 lb. in weight, was intended for killing rooks and rabbits, especially the former, and was sold by the gunmakers of its day for 12 to 15 gs.

It was contemporary with the improved air-gun with a hollow stock which superseded the air-gun that held the condensed air in a metal ball attached below its barrel.[1]

It is true the bullet crossbow did not shoot with the force of an air-gun, but it answered its purpose and was easier and safer to manipulate than any air-gun.

After the introduction of small rifles for shooting rooks—about 1840— bullet crossbows and air-guns were laid aside, though many of the former passed into the hands of poachers, who, owing to their silent discharge, found them useful for killing pheasants at roost.

Considerable amusement may, however, be derived from a good bullet crossbow, whether in knocking young rooks off the branches of not too tall trees—which it will do well—or in practising at a mark.

These weapons may be discovered—nearly always without their bow-strings —in the shops of provincial gunmakers and in those of dealers in curiosities, and often in the gun-rooms of old country houses which stand near rookeries. Few people are aware how well and truly they were made, how accurately they shot or how much they were valued by sportsmen in former days.

---

[1] The air-gun was invented in 1560 by Guter of Nuremberg. It was sometimes used in warfare in the first half of the eighteenth century, one German regiment of infantry even carrying these weapons instead of fire-locks.

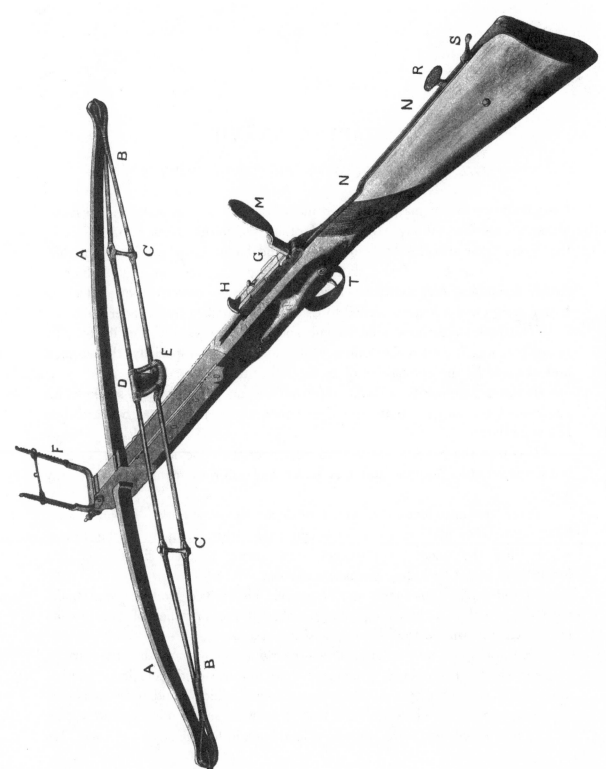

FIG. 115.—THE BULLET CROSSBOW.

This crossbow has not, to my knowledge, ever been properly described, though both Daniel and Blaine in their books on rural sports notice it in a cursory manner, the last-named author giving a fairly good engraving of one.

Thomas Waring, in his 'Treatise on Archery,' 1824, also gives a small sketch and a short account of the weapon.

The improved crossbow in question appears to have been of English manufacture only. I can find no trace of its having been made or used abroad, though its predecessor the stonebow was a popular sporting weapon in France, Germany and Italy, in the sixteenth and seventeenth centuries. The modern Continental bullet crossbow has a barrel, Chapter XLVI.

I give below the names of some of the best makers in former days of bullet crossbows such as here described:

R. Braggs, 43 High Holborn, London; T. Jackson, 29 Edward Street, Portman Square, London; Gameson & Co., London; Parker, Bury St. Edmunds; Barker, Wigan; J. Johnson, Manchester; Hyham, Warrington.

Hyham (now Daintith) was famed for the excellence of his crossbows and, I believe, made a larger number than anyone else. The gunmakers of Chester were also well known for the powerful and accurate bullet crossbows they produced.

I will explain the construction and management of this weapon for the benefit of those of my readers who may happen to possess one that is out of repair, and who wish to put it into working order.

FIG. 115, OPPOSITE PAGE, IS A BULLET CROSSBOW. ITS PARTS ARE:

A.A. The steel bow (length about 2 ft. 6 in.; width at centre, $\frac{3}{4}$ in.; thickness at centre, $\frac{5}{8}$ in.).

B.B. The bow-string. C.C. The cross-trees. D. The pocket for the bullet. E. The loop behind the pocket which is hitched by hand over the catch of the lock preparatory to bending the bow.

F. The metal fork across which the skein of thread is stretched that carries the bead which acts as a fore-sight. This fork hinges down flat when not required for taking aim.

G. The lock and its case.

H. The catch of the lock which, when the bow is bent, holds the stretched string secure till it is released by pulling the trigger T, to be seen beneath the stock.

M. The back-sight with its peep-holes. This, like the fore-sight, hinges down flat (over the top of the lock) when not required for aiming.

N.N. The steel lever. This lever pulls back the lock together with the

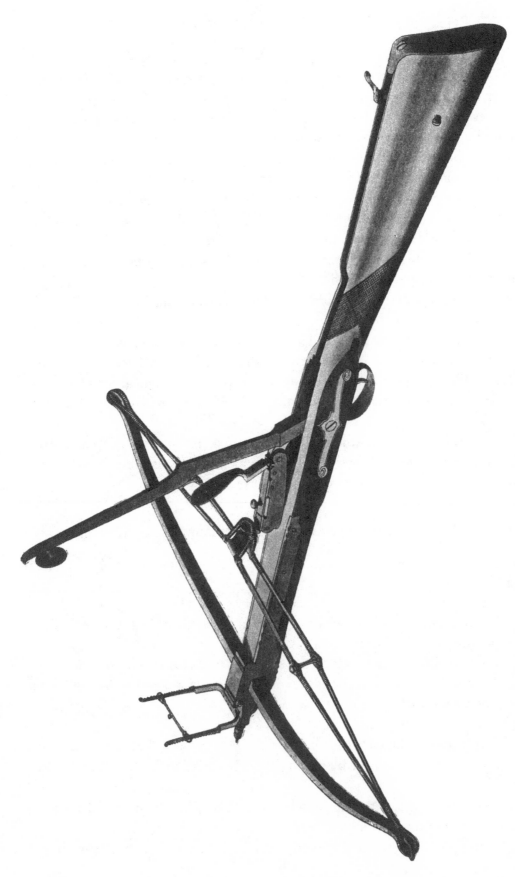

FIG. 116.—THE STRING ON THE CATCH OF THE LOCK AND THE LEVER HINGED FORWARD READY TO BEND THE BOW.

bow-string, when the latter is hitched over the catch of the lock. The lever is hinged to the stock and also to the casing of the lock, fig. 116, opposite page. It fits into its recess in the butt-end of the stock when the string of the crossbow has been stretched, or when the weapon is not in use.

R. The knob fixed to the top surface of the loose end of the stringing lever. By means of this knob the right hand presses the lever towards and finally into its recess in the stock, as the bow-string is being stretched and the bow bent, fig. 117, next page.

S. The spring thumb-catch which secures the end of the lever in the stock when the string of the crossbow is fully stretched. This catch also releases the end of the lever from the stock, so that it may be hinged forward preparatory to stretching the string of the crossbow again after the weapon has been discharged, fig. 116.

HOW TO BEND THE BOW AND STRETCH THE BOW-STRING

(I) In fig. 116, opposite page, the stringing lever is hinged forward out of its recess in the stock. The loop—which is behind the pocket of the bow-string—is hitched (by hand) over the catch of the lock, and the bullet should be in position in its leather pocket.

(II) Hold the butt of the stock firmly in the hollow of the left hand, the crossbow directed downwards, with its stock near and partly across the left side of the body. Fig. 117, next page.

(III) Place the palm of the right hand over the large knob of the lever. Press the knob and lever together (fig. 117, next page), steadily upwards towards the butt-end of the stock, till the loose end of the lever snaps into the notch of the spring thumb-catch and is thus safely secured, as shown in fig. 118, p. 183.

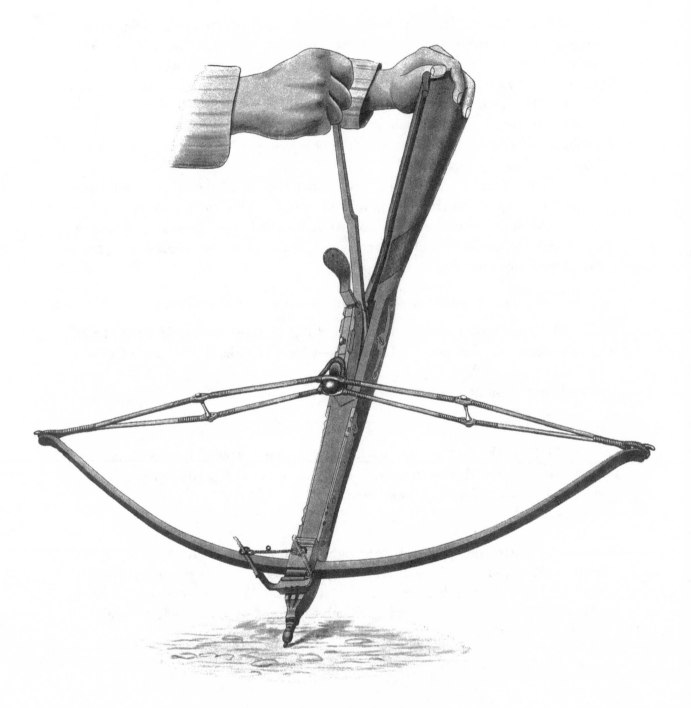

FIG. 117.—BENDING THE CROSSBOW.

FIG. 118.—THE CROSSBOW WITH ITS BOW BENT AND READY TO BE AIMED AND DISCHARGED.

CHAPTER XXXVIII

THE BULLET-SHOOTING CROSSBOW—ENGLISH
(*Continued*)

THE BASTARD-STRING

As the bow-string of a bullet crossbow is the part of the weapon that is usually in worst repair, it most needs description.

How to make the bow-string of a crossbow of this kind has long been forgotten. I am not, indeed, aware of anyone except myself who can now make and fit one properly. For this reason there are many fine bullet crossbows without bow-strings which might otherwise be a source of much amusement to their owners.

Before describing the bow-string of a bullet crossbow, I must explain the construction and application of its bastard string, for without the assistance of the latter its bow-string could not be fitted.

The bastard or false string of a crossbow was a necessary part of its equipment, though no doubt in the case of a company of crossbowmen one bastard string formerly served many crossbows, when new bow-strings had to be fitted to them or when old ones had to be removed for repair.

The bastard string was used to bend the steel bow of a crossbow sufficiently to allow its proper bow-string to be fitted on the bow. As the bow-string was always shorter than the bow, the latter had, of course, to be bent to a certain degree to enable its string to be put in position.

The bow of a crossbow, when its string was fitted, was always slightly bent so that its string might fit tight and true.

In the case of a wooden bow such as a longbow, it

FIG. 119.—THE BASTARD STRING AND ITS CLAMPS.

is perforce unstrung when laid aside, otherwise it would soon lose its elasticity, and hence its power.

The steel bow of a crossbow was, however, invariably forged with a curved outline. This natural bend prevented the bow from taking a 'set' when it was additionally bent to fit its bow-string, even though the strain thus entailed on it was a permanent one.

Without the mechanical assistance of a bastard string it would be impossible to bend a fairly powerful steel bow enough to put on its bow-string, if the latter is to fit as tight as it should do.

As the bow-string of a crossbow has so short a draw, it requires to start from a state of considerable tension when pulled back by its lever.

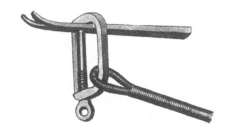

Fig. 120 shows the side and end view of one of the pair of screw clamps which, combined with the bastard string attached to them, were employed to bend the bow of a bullet crossbow to fit on its bow-string.

FIG. 120.   FIG. 121.

THE CLAMPS FOR THE BASTARD STRING OF THE ENGLISH BULLET CROSSBOW.

Fig. 121 represents one of the clamps screwed fast to one end of the bow, with one end of the bastard string attached to it.

The bastard string, formed of a skein (about $\frac{3}{8}$ in. thick) of strong pack-thread (twisted cord is useless for the purpose, as it stretches under pressure), is wrapped with whip-cord at its ends and centre to keep its strands together. The bastard string should stretch from one clamp to the other when these are fixed to the bow, fig. 119, opposite page.

To bend the bow, preparatory to putting on or taking off the bow-string, the centre of the bastard string is hitched by hand over the catch of the lock of the crossbow, the lever being hinged forward out of the stock for this purpose, as shown in fig. 122, next page.

The lever is then pulled back till the bow is sufficiently bent to allow

Fig. 122.—The Crossbow bent sufficiently by the Bastard String for the Skein (or Bow-string) to be fitted to it.

the end loops of the bow-string to be slipped over the points of the bow into the notches (or out of them, should you wish to remove the string) shaped to receive them, fig. 122, opposite page.

When the string is being put on the bow, its looped ends are passed through the clamps, fig. 126, p. 190. The bow-string being finally placed on the bow, the screws that fix the clamps are unturned. The clamps then being loose can be removed from the bow over its ends, and they and the bastard string laid aside till again required.

If the bow of a bullet crossbow is of weak construction, the bastard string need merely consist of a skein of fine twine of a thickness of $\frac{1}{4}$ in. The skein should have a loop knotted at each end, and these loops should be tightly lashed with waxed thread to the bow, at points respectively 2 in. short of its opposite ends.

FIG. 123.—How to fix a Bastard String to a Light Steel Bow.

Each loop will need a small wedge of hard wood driven under it to tighten it, (fig. 123,) or it will slip along the smooth surface of the bow when pressure is applied by the lever to bend the latter and fit the bow-string.

When the bow-string is finished and fitted, the bastard string is, of course, taken off the bow.

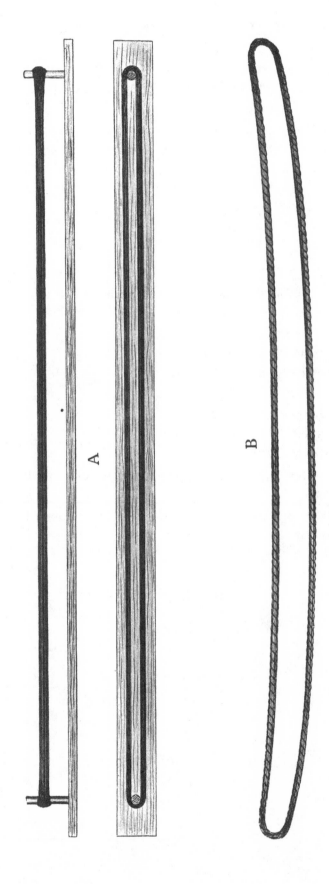

FIG. 124.—THE SKEIN OF THE BOW-STRING.

CHAPTER XXXIX

THE BULLET-SHOOTING CROSSBOW—ENGLISH (*Continued*)

HOW TO MAKE AND FIT THE BOW-STRING

A, fig. 124, opposite page, is the skein of fine twine, side and surface view—bookbinder's twine for preference, then the strong brown packthread employed for sewing carpets and other thick materials. The twine should be wound—about 120 times—round and round two smooth pegs of wood (each $\frac{1}{2}$ in. in diameter) driven upright into holes in a board. The pegs should be fixed with their outer edges at a distance from one another that is $\frac{1}{2}$ in. shorter than the distance between the notches at the ends of the steel bow.

FIG. 125.—ONE END OF THE SKEIN ON ITS PEG. SIDE AND SURFACE VIEW.
Half full size.

Each half of the skein when its threads are pressed together, should be about as thick as an ordinary lead pencil, fig. 125.

———————

B, fig. 124, shows the skein as it appears after it has been removed from the pegs.

———————

The outside wrapping of fine twine to be seen on the skein, (B, fig. 124,) holds its threads together when it is taken off the pegs and prevents it falling into a tangle.   This outside wrapping should be done in turns $\frac{1}{8}$ in. apart before the skein is lifted off the pegs.

FIG. 126.—FITTING THE SKEIN OVER THE ENDS OF THE BOW WHEN THE LATTER IS SUFFICIENTLY BENT BY THE BASTARD STRING FOR THIS PURPOSE (SEE ALSO FIG. 122, P. 186).

Secure the stock of the crossbow in a bench-vice, then by means of the bastard string and the application of the lever of the crossbow, bend the steel bow till the ends of the skein can be slipped into the notches at the ends of the bow, as shown in fig. 126.   How to apply the bastard string for this purpose is explained in Chapter XXXVIII.

The cross-trees should now be fixed to the skein.

These consist of two little pillars of ivory turned in a lathe, their ends being notched to prevent them from slipping when fitted.   Each pillar is $\frac{1}{4}$ in. thick by $1\frac{5}{8}$ in. long ; one is shown in fig. 127.

FIG. 127.—A CROSS-TREE.   Full size.

Place a cross-tree through both halves of the skein, at a point that is 6 in. from the centre of the latter.   Place the other cross-tree in a corresponding position, c c, fig. 115, p. 178, and fig. 137, p. 193.

The ends of the cross-trees should divide equally the threads of the skein where they pass through them.   On each side of the cross-trees the skein—or bow-string as it may now be called—will have to be wrapped tightly round with fine waxed whip-cord, for a length of 1 in. One of the cross-trees as fixed in the bow-string, is shown in fig. 128.

FIG. 128.—ONE OF THE CROSS-TREES IN POSITION IN THE BOW-STRING.
Half full size.

Next, slightly bend the bow again with the bastard string and the lever to remove the bow-string, then tightly wrap each end of the bow-string with waxed whip-cord, as in fig. 129.

FIG. 129.—ONE OF THE ENDS OF THE BOW-STRING. Half full size.

By the aid of the bastard string the bow-string may now be replaced on the bow—this time permanently. When this is done, the bastard string and its clamps can be removed as they are no longer required.

The pocket (formerly called from its shape the cradle) to hold the bullet comes next and requires to be made strongly and neatly.

(1) Hold a piece of round wood, 1 in. in diameter and 3 in. long, in an upright position against the centre of the bow-string and on that side of it which is towards the butt-end of the crossbow. The centre of the bow-string should, of course, be above the centre of the stock, fig. 130.

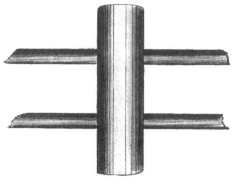

FIG. 130.—THE PIECE OF ROUND WOOD AS HELD BY THE FINGERS AGAINST THE BOW-STRING. SIDE AND SURFACE VIEW. Half full size.

(2) Whilst you hold this piece of wood against the bow-string, wrap fine twine (the same as that of which the bow-string is made) to and fro round the wood and over and under each half of the bow-string, till you have formed two separate loops round the wood each about the thickness of a lead pencil, fig. 131.

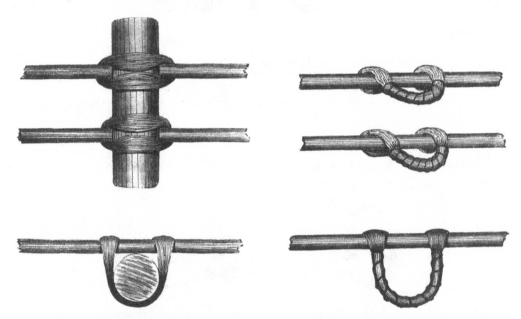

FIG. 131.—THE PIECE OF WOOD WITH THE LOOPS FORMED ON IT. SIDE AND SURFACE VIEW. Half full size.

FIG. 132.—THE PIECE OF WOOD TAKEN AWAY AND THE LOOPS WRAPPED WITH SILK. SIDE AND SURFACE VIEW. Half full size.

(3) Without shifting the piece of wood from the bow-string, wrap the loops closely round with some soft silk to hold their strands together when

the piece of wood is removed. A curved needle will enable you to pass the silk round the loops where these encircle the piece of wood. When this is done, the piece of wood may be taken away, fig. 132.

(4) Without undoing the silk wrapped round the loops, wrap each loop throughout with fine waxed whip-cord, fig. 133.

FIG. 133.—THE LOOPS WRAPPED WITH WHIP-CORD. SIDE AND SURFACE VIEW. Half full size.

(5) Next, and with a slightly coarser whip-cord, also well waxed, lash the two loops together for an inch in length at their centres, so as to make them

true and solid at the part where they are jointly hitched over the catch of the lock.

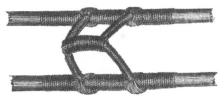

FIG. 134.—THE LOOP FINISHED.
Half full size.

Between the four ends of the loop, where these surround the bow-string, and for an inch on each side of them, wrap the bow-string with fine waxed whip-cord.

Fig. 134 shows the loops or loop as it now is, finished, together with the wrapping of whip-cord along the bow-string.

(7) Sew a strip of leather ($\frac{3}{4}$ in. wide and 3 in. long), between the four ends of the loop, the ends of the piece of leather to be turned back and sewn round each half of the bow-string. This strip of leather should be a little loose when fixed, so that it may partly enfold the $\frac{1}{2}$ oz. bullet which the crossbow discharges and thus prevent it from falling out of its pocket, figs. 135, 136.

The bullet is placed in this pocket after the loop of the bow-string has been

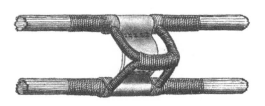

FIG. 135.—THE LEATHER POCKET FITTED TO THE LOOP. FRONT AND BACK VIEW.
Half full size.

FIG. 136.—SECTION OF LOOP AND LEATHER POCKET, WITH BULLET IN POSITION IN THE POCKET.
Half full size.

hitched by hand over the catch of the lock, preparatory to applying the stringing lever to bend the bow, fig. 116, p. 180, and fig. 136.

When the loop behind the pocket is hitched over the catch of the lock, the slight strain caused thereby brings the halves of the bow-string at its centre somewhat together, with the result that the leather pocket grasps the bullet and holds it securely, as shown in fig. 136.

FIG. 137.—THE BOW-STRING AS IT SHOULD APPEAR WHEN FINISHED. FRONT VIEW.

Note.—If the finished bow-string is a trifle too tight—which may be known by the pocket for the bullet inclining slightly downwards towards the stock, instead of pointing straight forward as it should do—you can slacken it a little by wrapping round it a cloth soaked in boiling water, meanwhile stretching it several times with the stringing lever.

On the other hand, if the bow-string is rather loose—instead of taut—remove it by means of the bastard string in the manner previously explained. Then fit over the pointed ends of the bow (formerly called from their shape ' the thumbs ') two thick leather washers each about the size of a shilling, with holes, of course, in their centres. As these washers will rest between the notches of the bow and the end loops of the bow-string, they will have the same effect as shortening the latter.

CHAPTER XL

THE BULLET-SHOOTING CROSSBOW—ENGLISH (Concluded)

THE LOCK AND THE SIGHTS

In a bullet crossbow, the lock and its casing are hinged to the metal lever ($\frac{5}{16}$ in. thick), which bends the steel bow, fig. 122, p. 186.

The working parts of the lock consist of a tumbler, sear, lock-trigger and spring. They are fitted between two steel side-plates, each about $\frac{1}{8}$ in. thick. These side-plates form the casing of the lock.

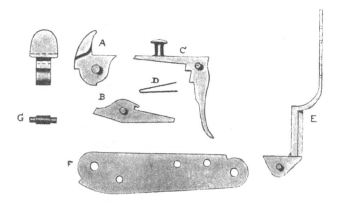

FIG. 138.—THE PIECES OF THE LOCK. Half full size.

A, B, C, D, E, are each $\frac{5}{16}$ in. wide transversely where they work between the side-plates.

A. The tumbler, front and side view ; its hook-shaped upper part forms the catch which holds the bow-string.

B. The sear.   C. The lock-trigger.

D. The lock-spring.   E. The peep-sight.

F. One of the side-plates.

G. One of the step-ended rivets that hold the side-plates together and on which the pieces of the lock are hinged.

A, fig. 139.   The lock is here to be seen in its normal or 'set' condition, with its catch in position to hold the loop of the bow-string.

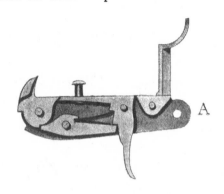

The curved catch of the lock, *i.e.* the upper part of the tumbler (A, fig. 138, previous page), stands above the side-plates at the forward end of the casing. This catch is smoothly rounded at its edges so that it may hold the bow-string without cutting it.

The upright iron at the lever end of the lock is the base of the peep-sight (M, fig 115, p. 178). As this piece has no connection with the movements of the lock, its upper part is here omitted.

FIG. 139.—THE LOCK WITH ITS WORKING AND OTHER PARTS FITTED, AND ONE OF THE SIDE-PLATES OF ITS CASING REMOVED TO SHOW THE INTERIOR ACTION OF THE LOCK.

Half full size.

B, fig. 139.   The lock as it appears when the trigger (T, fig. 115, p. 178) in the stock of the cross-bow has been pulled to release the bow-string.

This trigger, which swings loosely in the stock of the crossbow, presses back the projecting end of the lock-trigger. This allows the long end of the sear to escape from the little step in the lock-trigger against which it rested, as it did in A, fig. 139. At the same moment, the strain of the bow-string on the catch of the tumbler causes the tumbler and the short end of the sear to disengage where they were previously interlocked, the loop of the bow-string being, of course, then instantly set free from the catch.

It will be seen that the short end of the sear and the notch in the tumbler are slightly sloped (fig. 138, previous page), where they fit against one another when the lock is set, as it is in A, fig. 139. This causes these parts to separate directly the long end of the sear and the trigger of the lock are clear of each other. To reset the lock, all you need do is to push the tumbler upwards into the position shown in A, fig. 139.

The shoulder in the tumbler, which may be seen immediately below its upper part or catch, comes against the short end of the sear when the lock is set free and thus prevents the tumbler from turning too far round.

By pressing down the small knob fixed on the top part of the lock-trigger, it can be utilised to free the bow-string.

Sometimes when the bow-string has been hitched by hand over the catch of the lock preparatory to bending the bow with the lever (fig. 116, p. 180), the intention of doing so is changed.   As the trigger in the stock can only be worked when the bow is fully bent, this knob is often useful in such a case for instantly loosing the string from the catch of the lock.   The trouble of lifting it off the catch is thus avoided.

In many of the older bullet crossbows the trigger in the stock was absent, this small knob taking its place as the only means of setting free the bow-string when the bow was fully bent.

THE SIGHTS OF THE BULLET CROSSBOW, FIGS. 141, 142, NEXT PAGE

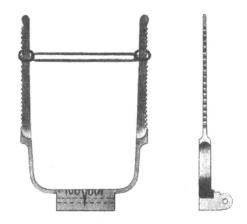

FIG. 140.—THE FRAME OF THE FORE-SIGHT, WITH ITS SMALL SKEIN OF THREAD STRETCHED BETWEEN TWO METAL RINGS.   Half full size.

The frame that holds the bead which acts as a fore-sight is the notched steel fork attached to the fore-end of the stock, F, fig. 115, p. 178.   This fork hinges down when not required.

To arrange the fore-sight :

(1) Obtain a couple of metal rings $\frac{1}{4}$ in. in diameter, such, for instance, as are used for the top joint of a salmon rod.

(2) Place one ring on one arm of the notched fork and the other ring opposite to it on the other arm of the fork.

(3) With a needle and a couple of feet of black sewing cotton, connect the two rings together, so as to form a tight little skein (of about a dozen strands) between the upright arms of the fork, fig. 140.

(4) Take another length of black cotton and wrap it closely over this little skein and to within $\frac{1}{4}$ in. of each ring.

(5) When you arrive at the centre of this last wrapping, thread a small white bead on the cotton you are twisting round the skein. Leave the bead threaded above the centre of the skein and proceed to finish the other half of the wrapping, fig. 141.

FIG. 141.—THE FORE-SIGHT, WITH THE SIGHTING BEAD ON ITS SKEIN.

Half full size.

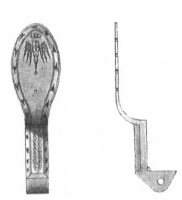

FIG. 142.—THE BACK-SIGHT, *i.e.* PEEP-SIGHT.

Half full size.

The back-sight, fig. 142. This has four or five peep-holes to suit the different distances at which the crossbow is used. The peep-holes are enlarged in the usual way on the reverse side of the sight to that through which aim is taken.

HOW TO AIM THE CROSSBOW

Fasten up against a wall at 25 yards distance, a large sheet of white paper marked with a black bull's-eye 3 in. in diameter, the 'bull's-eye' being level with your eye.

1. See that the little rings on the arms of the fore-sight are in opposite notches and rather more than halfway up the fork, so that the bullet may pass below the skein of thread without any risk of striking it when the crossbow is discharged.

2. Look through the central peep-hole in the back-sight, the other peep-holes being temporarily stopped with beeswax.

When you see through the one open hole in the back-sight that the bead of the fore-sight covers the centre of the bull's-eye, pull the trigger of the crossbow.

3. If the bullet strikes a trifle low but straight beneath the bull's-eye, lower the rings of the thread skein two or three notches down the arms of the fork.

4. If the bullet strikes a trifle high and straight above the bull's-eye, raise each ring of the skein two or three notches up the fork.

5. Should the bullet go to the left, move the sighting bead a trifle along the skein to the left by revolving it in that direction round the cotton on which it is threaded.

6. Should the bullet go to the right, twist the bead along the skein a few turns to the right.

7. If the bullet strikes a good deal too high or too low, look through a peep-hole in the back-sight which is higher or lower than the one you have been using, in order to acquire a proper elevation. A shot too low will be corrected by looking through one of the higher peep-holes in the back-sight, and a shot too high by the use of one of the lower ones.

When you have accurately sighted the crossbow for a range of 25 yards, which is far enough for ordinary rook-shooting and as far as the bow will kill with certainty, fix the sights beyond alteration.

To do this, wrap a little fine waxed silk on each side of the sighting bead to keep it from being accidentally moved, and fill in with beeswax all the peep-holes in the back-sight except that which you find is the correct one to aim through.

When the sighting mechanism and string of the bullet crossbow are properly adjusted, you should be able to hit a playing card eight times out of ten at from 20 to 25 yards.

I have many times brought down six rooks from the tops of fairly tall trees in six consecutive shots with one of these weapons. The absence of all noise on the part of the crossbow, will allow you to go quietly into a rookery and kill a number of young birds before the old ones become alarmed.

If held at an angle of 45 degrees, a good bullet crossbow will throw a $\frac{1}{2}$ oz. lead bullet to an extreme range of 300 yards, and if shot at a metal target at 20 yards, more than half of the bullet will be flattened. The weapon can easily be made ready and then aimed and discharged, four times in a minute.

To preserve the bow-string of the crossbow, be sure to keep it well rubbed with beeswax. At any part where the string is inclined to fray, wind some waxed silk tightly round to keep it together. If the bow-string is properly made it should last for a score of years in frequent use.

The bow of a bullet crossbow being comparatively slight and much bent when its string is stretched, it should never be kept in this condition for longer than a few minutes at a time. It is better to discharge the weapon into the ground (you then save the bullet), than to keep its bow too long in a state of tension and thus run the risk of straining it.

The best method of taking about the crossbow when chances of shots are occasional, is to carry it with its bow-string hitched over the catch of the lock, the bullet in the pocket of the string and the lever extended.   See fig. 116, p. 180.   Then when a shot offers, the lever can be at once pressed home into the stock and the weapon is instantly made ready for use.

With a little practice, the lever can be set free from its catch and the bow-string in this way slackened to unbend the bow if it is not likely to be discharged for some time.

To do this, hold the crossbow in your left hand as when using the lever to bend its bow, p. 181.   Place the right hand firmly over the knob of the lever and at the same time pull back with the left thumb, (*i.e.* the thumb of the hand that grasps the butt,) the spring catch which secures the lever in its cavity in the stock.   This will enable you to let the lever come gently forward out of the stock.

In this manner you can gradually unbend the bow, the action of doing so being the reverse of that shown in fig. 117, p. 182.

CHAPTER XLI

THE LARGE BOLT-SHOOTING CONTINENTAL TARGET CROSSBOW

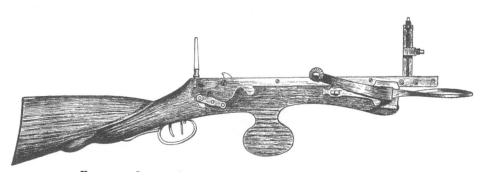

FIG. 143.—LARGE CONTINENTAL TARGET CROSSBOW. SIDE VIEW.

THIS crossbow, which may be considered in some measure a revival of the mediæval weapon, was of admirable design and construction and had as powerful a steel bow as it was possible to bend with a goat's-foot lever.[1]

Its bow was bent as shown in the crossbow, fig. 45, p. 89, and as is also represented at the end of this Chapter. Its goat's-foot lever was, of course, of a size proportionate to the strength of the bow and to the distance the bow-string had to be drawn along the stock of the crossbow.

The lock had the two triggers commonly seen in all the best sporting and target crossbows made in the latter half of the sixteenth century, the back trigger being employed to cock the lock, and the front-, or hair-trigger to discharge the crossbow.

The circular steel catch for the bow-string was the same in shape as the ivory nut of mediæval times. The catch and the lock of this crossbow were identical with the catch and lock shown in fig. 113, p. 173.

The transverse metal pin for the goat's-foot lever had its ends fitted with thin collars of steel. These revolving collars assisted the downward slide of the arms of the lever when the latter was being used to pull back the bow-string.

[1] See Chapter XVII, for a description of the goat's-foot lever.

Fig 144.—Large Continental Target Crossbow.

The fore-sight of this crossbow was very ingenious. Its sighting bead or point could be elevated or depressed to suit the range at which the crossbowman desired to shoot. It could also be moved to the right or left, to allow for a side wind or to alter the line of flight of the bolt when it did not travel straight, fig. 145.

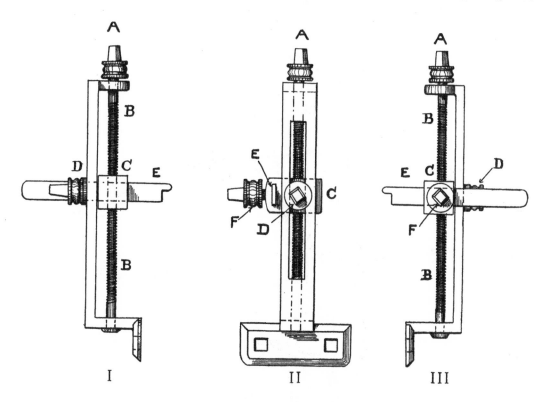

FIG. 145.—THE FORE-SIGHT. Full size.

I and III, reverse views. II, side view.

A The squared top of the upright screw-rod B, B. By turning A, one way or the other, the screw-rod moves the block C (through which it passes), up or down the frame of the fore-sight.

E The fore-sight bar. The sighting bead or notched point of this bar projects over the groove of the stock at the fore-end of the crossbow. The bolt passes beneath this point when the weapon is discharged, fig. 144.

It will be seen that the fore-sight bar E, and the block C, travel together up or down the screw-rod when the latter is revolved by turning A.

By untwisting the screw F, the fore-sight bar E can be pushed to or fro in its hole in the block C, till its point is in a position to suit the aim ; then by re-tightening F, the bar can be fixed.

The screw D, travels with the block C, up or down the upright opening in the framework. This screw D, is used to clamp C, when the latter is at a correct

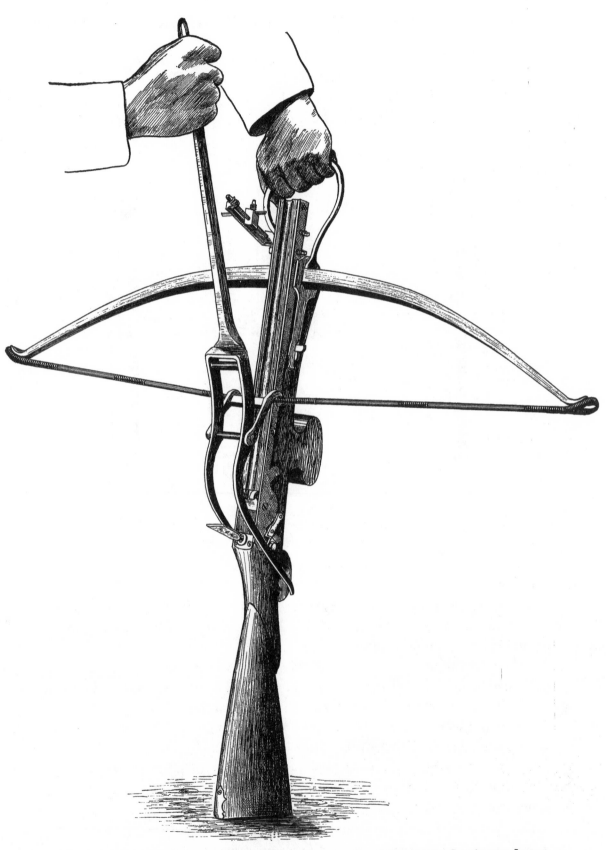

FIG. 146.—LARGE CONTINENTAL TARGET CROSSBOW BEING BENT BY ITS GOAT'S-FOOT LEVER.

height as regards the elevation of the fore-sight bar E above the stock of the cross-bow. The screw-heads of the frame are adjusted by means of a little spanner.

The back-sight is an ordinary peep-sight. Aim is taken by first looking through one of the apertures in the back-sight, and then covering the mark with the small point E, of the fore-sight bar.

The stock of the crossbow is 2 ft. 6 in. long, and $1\frac{1}{2}$ in. wide across its grooved surface. From its fore-end to the catch of the lock 15 in., the draw of the bow-string being 6 in.

The steel bow is 3 ft. long. It is $\frac{7}{8}$ in. wide and $\frac{3}{4}$ in. thick at its centre.

The bolt is 12 in. long and $\frac{9}{16}$ in. in diameter, and weighs $1\frac{3}{4}$ oz. In shape it is the same as the bolt shown in fig. 157, p. 218.

The target-shooting range of this crossbow is from 70 to 80 yards, its extreme range about 280 yards.

These weapons were very popular on the Continent, especially in Germany, Switzerland and Belgium, from about 1750 to about 1820.

It is said that in the first quarter of the eighteenth century similar crossbows were employed by the Swiss and Tyrolese hunters for killing chamois.

It is possible an earlier weapon of the kind was thus used,[1] though certainly not the one shown opposite, which, owing to its delicate sights and high finish, was evidently intended for target practice only.

There is no doubt that at 60 yards, a sharp heavy bolt discharged from a crossbow with a steel bow of such length and strength, would cause the death of a chamois. It should also be remembered that there were a hundred chamois to shoot at in the beginning of the eighteenth century where there would be only one now.

[1] How to stalk and kill chamois with the crossbow is fully told by Gaston Phœbus in his famous book on sport written in the fourteenth century (see note on Gaston Phœbus, p. 78). I may add that the sporting crossbow of the fourteenth century was not so powerful or accurate a weapon as the one described in this chapter.

CHAPTER XLII

THE SMALL BOLT-SHOOTING TARGET CROSSBOW AS NOW USED IN BELGIUM

HERE we have a crossbow which somewhat resembles that described in the last chapter, though in this case the steel bow is smaller and is not bent by a goat's-foot lever.

This Belgian weapon as now made, represents the experience of centuries of cross-bow construction, and is the most perfect article of its kind at present produced.

For nearly three hundred years, this form of crossbow has been popular for target-shooting in the north of France and in Belgium.[1]

Its chief place of manufacture and sale is Brussels, where several firms deal in these weapons.

In Brussels, Ghent, Bruges and Antwerp and in part of the north of France, and their environs, there are to this day societies of crossbowmen, whose members compete for prizes at the ranges attached to inns, clubs and private residences.

The Belgian target crossbow, though it has a small bow in comparison with what was used in mediæval days, shoots with wonderful accuracy and considerable power.

With this crossbow, I find that at a distance of 60 yards, I can generally place eight out of twelve bolts in the 6-in. centre of the usual archery target, and the other four bolts close round the edge of the gold.

At a range of 60 yards, the weapon sends its bolts with a force that causes them to pass through the ordinary straw target and often out of sight beneath the ground beyond.

As this spoils the target, the bolts being so much thicker than the arrows of a longbow, it is advisable to pack between its painted face and the straw behind it, three or four layers of stout canvas.

The extreme range of the Belgian crossbow is about 250 yards. It is almost as accurate as a rook rifle up to 50 yards, and its mechanism, including its lock, sights and hair-trigger, is excellent.

---

[1] Shooting with the crossbow is a favourite pastime in Saxony, particularly in the vicinity of Dresden, Chapter XLVIII. Crossbow-shooting is also practised to some extent in Switzerland and in Holland.

The steel bow is powerful for its length, and of superior finish and material.

This crossbow is well adapted for target practice in the grounds of a country house. It is capable of affording much amusement, and is noiseless, safe and easy to manipulate.

It is, however, more than a toy, as at a range of 60 yards it has sufficient strength to drive its bolt through a hare or rabbit.

For this reason, should the crossbow be carried on a summer's evening for killing rabbits whilst feeding it is best to employ bolts with blunt heads instead of the sharply pointed ones that are used at the target. The bolts with sharp points are liable to be lost if discharged at rabbits, as they are apt, if they miss the mark, to penetrate out of sight beneath the ground.

I will describe this crossbow in detail, as it is one that an amateur mechanic of fair skill should be able to make.

If the lock and its catch present difficulties, the simple mediæval lock given in Chapter XXI may be substituted, which, though not so suitable for accurate shooting as the one shown in Chapter XLIII, will answer fairly well.

To obtain a steel bow of correct size and shape, first cut out an exact model of it in wood, then send the model to a spring maker or better still to a Liège gunmaker, to have it reproduced in finely tempered steel.

Do not forget that the ends of the bow should be canted up $\frac{1}{2}$ in. above its centre, as explained in fig. 58, p. 102.

If the ends of the bow are not given this upward cant, the bow-string will press too hard upon the stock of the crossbow, with the result that the friction of the bow-string, as it travels along the surface of the stock, will greatly reduce the velocity of the bolt.

It might be thought that the stock of this cross-bow was unduly large in proportion to the size of its bow.

A fairly large and heavy stock is an advantage in a crossbow used at the target, as it gives steadiness to the aim and accuracy to the bolt. The projecting curved handle below the stock is grasped by the left hand of the crossbowman when he is in the act of aiming, and enables him to hold the weapon securely in a level position as he pulls its trigger.

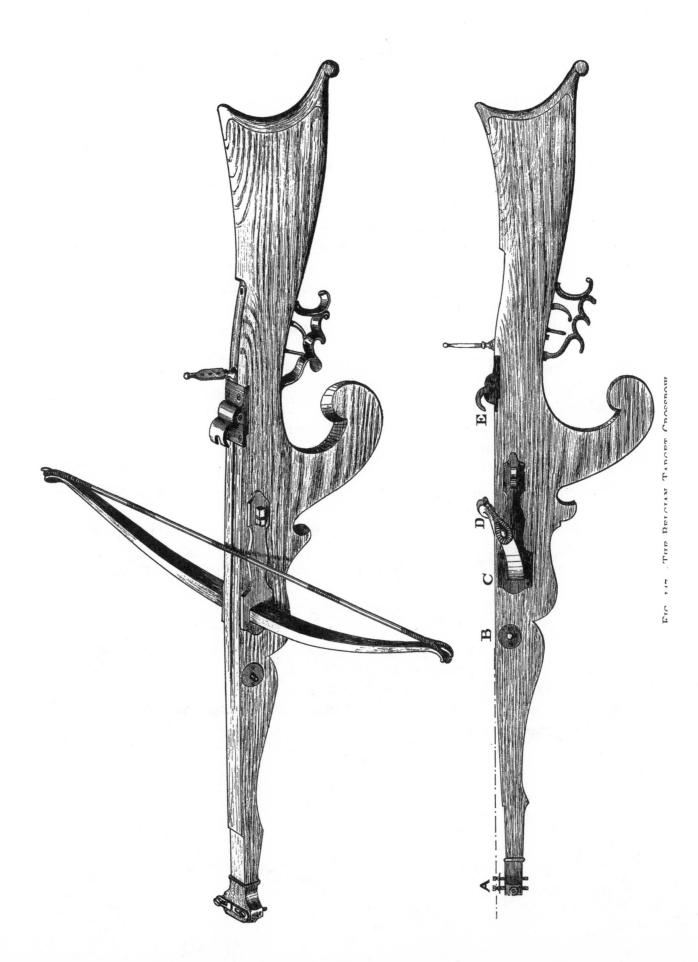

FIG. 147.—THE BELGIAN TARGET CROSSBOW.

DIMENSIONS OF THE BELGIAN TARGET CROSSBOW
FIG. 147, OPPOSITE PAGE

THE STOCK

	ft.	in.
Total length of the stock.	3	10
Thickness of the stock	0	$1\frac{3}{8}$
Depth of the stock where the bow is fixed . . .	0	3
Depth of the stock at its fore-end	0	1
From the fore-end of the stock to the back of the bow, A–C	1	6
From the transverse metal pin (for the fork of the lever) to the back of the bow, B–C	0	3
Drop of the surface of the stock at its fore-end A, below the point B	0	$0\frac{3}{4}$
Length of the groove in which the bolt is laid—starting from the catch for the bow-string—E, B . . .	1	0
From the inside of the bow to the catch for the bow-string, C–E	0	9
Draw of the bow-string, from a state of rest to the catch of the lock, D, E	0	$5\frac{1}{2}$

THE STEEL BOW

	ft.	in.
Total length of the bow	2	3
Width of the bow at its centre of length . . .	0	1
Thickness of the bow at its centre of length . .	0	$0\frac{1}{2}$
Width of the bow halfway between its centre of length and each of its ends	0	$0\frac{7}{8}$
Its thickness at these points	0	$0\frac{3}{8}$
Width of the bow at 2 in. from each end . . .	0	$0\frac{5}{8}$
Its thickness at these points	0	$0\frac{5}{16}$

CHAPTER XLIII

BELGIAN TARGET CROSSBOW (*Continued*)

THE LOCK

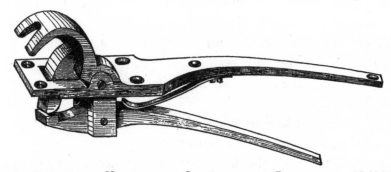

FIG. 148.—PERSPECTIVE VIEW OF THE CATCH FOR THE BOW-STRING. Half full size.

HERE we have a lock which resembles the one shown in figs. 108–112, pp. 170–172, except that in this case there is no notch across the surface of the stock to assist in holding the stretched bow-string.

This lock has the usual two triggers of the late mediæval crossbow, the back trigger being used to cock the lock and the front trigger to discharge the crossbow.

The fingers of the catch drop over the bow-string and hold it securely when the bow is bent. When the front trigger is pulled, the catch tilts upwards and thus allows the bow-string to escape from the grip of its fingers.

This is an excellent form of catch for the bow-string of a crossbow of moderate strength. Its action is identical with that of the fingers of the archer when he releases the string of his longbow.

―――――――――

It will be seen that the parts of the lock are very compactly arranged, so that they may be inserted in the stock of the crossbow with as little cutting away of its interior as possible.

THE MECHANISM OF THE LOCK, FIGS. 149, 150.

Fig. 149. A The catch of the lock, with its fingers tilted up ready to come down over the bow-string, see also fig. 148.

As the bow-string is forced along the stock by the stringing lever (fig. 156, p. 217), it is at length pushed hard against the projection B, which is beneath the fingers of the catch A and forms part of it, see also fig. 151, next page.

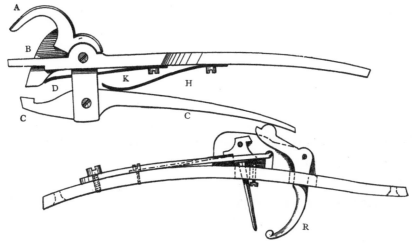

FIG. 149.—THE PARTS OF THE LOCK, AS FITTED INSIDE THE STOCK OF THE CROSSBOW.
Half full size.

This pressure of the bow-string against the projection B, causes its lower end to snap down into the notch near D, of the lever C C.

When B and C C are thus interlocked at D, the fingers of the catch will have come down over the bow-string E, as seen in fig. 150.

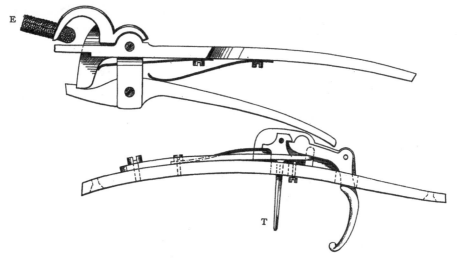

FIG. 150.—THE LOCK WITH THE FINGERS OF THE CATCH HOOKED OVER THE STRETCHED BOW-STRING E. Half full size.

When the small lock is cocked (as shown in fig. 150), by pulling back its curved trigger (R, fig. 149), the crossbow is ready for use.

By now pulling the front trigger (T, fig. 150), the hammer of the small lock is liberated, and at once knocks upwards the long end of the lever C C, see fig. 149.

FIG. 151.—THE CATCH SEPARATE FROM THE LOCK, AND THE PROJECTION B, THAT FORMS PART OF IT.

This causes the short end of the lever C, C, to drop clear of the end of B, at the notch near D, fig. 149.

As B is then free, the result is that the fingers of the catch instantly tilt up into the position seen in fig. 149, and allow the bow-string to escape from their grip.

FIG. 152.—SURFACE VIEW OF THE SMALL LOCK AND ITS SPRINGS.   Half full size.

The spring above H (fig. 149), presses down the longer half of the lever C C, and thus causes its short end to retain the lower end of B safely in the notch D, when the bow-string is hitched beneath the fingers of the catch, as it is in fig. 150.

The spring above K, by pressing underneath B, keeps the fingers of the catch always tilted up, ready for the bow-string to be pushed beneath them by the stringing lever, fig. 149.

This spring K, is of service in another way. When the end of B is interlocked with the lever C C, at the notch D, as in fig. 150, the spring K is then bent hard against the end of B. When B, through the action of the front trigger of the lock, is clear of the notch at D in the lever C C, then K, as it recoils from its bent position, instantly throws up the fingers of the catch and in this way gives a smooth and instantaneous loose to the bow-string.

CHAPTER XLIV

BELGIAN TARGET CROSSBOW (Continued)

THE SIGHTS

THE fore-sight (fig. 153), though here made entirely of metal, resembles in its principle that of the bullet crossbow, p. 198. It is also worked in the same manner.

I. The surface view of the fore-sight.

II. The end view.

III. The side view.

The fore-sight is hollow, and the fore-end of the woodwork of the stock fits into it. The fore-sight then forms the extremity of the crossbow, fig. 153.

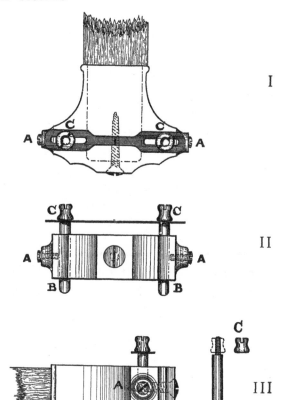

FIG. 153.—THE FORE-SIGHT. Half full size.

To move the sliding bar of the fore-sight to the right or left, loosen the screw-caps C, C, that fit on the tops of the small upright pillars, B, B. Then tighten these screw-caps again when the sighting bead on the centre of the bar is shifted into a position to suit the aim.

To alter the elevation of the sliding bar, loosen the side screws, A, A, which clamp the small upright pillars, B, B. Push the small pillars up or down, till the sliding bar (which connects their upper ends) is at the required height. Then secure the pillars from movement by re-tightening the screws A, A.

The sliding bar and its sighting bead, with a separate view of one of the pillars and its screw-cap, may be seen in fig. 153.

The back-sight is screwed into the top of the stock, 3 in. behind the catch for the bow-string.   It is an ordinary peep-sight, fig. 156, p. 217.

TO AIM THE CROSSBOW

In sighting the crossbow, be careful never to raise the sliding bar of the fore-sight more than $\frac{1}{2}$ in. above the surface of the metal beneath it, or the bolt is likely to strike the bar and smash it.

The height of the sliding bar need seldom be more than about $\frac{1}{4}$ in., II, fig. 153, previous page.

When the crossbow is discharged, its bolt is intended to pass close above the sighting bead on the centre of the sliding bar of the fore-sight.   The stock of the crossbow being sloped downwards to its fore-end, enables the bolt to leave the stock so that it passes above the bar of the fore-sight.   The dotted line, A–B, fig. 147, p. 208, shows the flight of the bolt as it leaves the crossbow.

If the bolt passed under the sliding bar, the bar would require to be so high above the stock to avoid contact with the bolt, that the latter would strike much below the mark at 40 yards, whatever the aim taken.

The groove for the bolt runs out to nothing from the part of the crossbow where its stock begins to slope downwards, and this part is 1 ft. 3 in. from its fore-end.

This arrangement of the groove (copied from weapons of mediæval days), causes the bolt to start off with a free and true flight, as it encounters no friction to divert its direction on quitting the stock of the crossbow.

When you have sighted the crossbow so that it will place eight out of twelve bolts in the gold of an archery target, at 50 yards distance, you may fix its fore-sight beyond movement, and stop up with beeswax all the peep-holes of the back-sight, except that which you find is the correct one to aim through.

If when sighting the crossbow at a mark, you find it sends its bolt too much to the right, you will have to move the sliding bar to the right till the bolt attains a straight course.   If the bolt inclines to the left, then the bar must be moved to the left.

The principle of sighting this weapon is identical with that of the bullet crossbow (pp. 198, 199).   The bead on the sliding bar, as seen through the peep-sight, gives the alignment for a correct aim.

CHAPTER XLV

BELGIAN TARGET CROSSBOW (Concluded)

THE LEVER, AND HOW TO USE IT TO BEND THE BOW

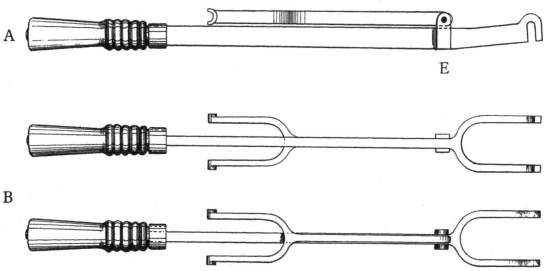

FIG. 154.—THE METAL LEVER FOR BENDING THE BOW.

Scale $\frac{1}{4}$ in. = 1 in.

A, Side view of the lever (folded up as when not in use).   B, Surface views of the lever (folded).

This powerful lever closely resembles the seventeenth century wooden lever shown in fig. 106, p. 167.

Length of the long arm of the lever, including its fork but without its wooden handle, 16 in.

The long arm is $\frac{3}{4}$ in. wide by $\frac{5}{16}$ in. thick ; its fork is $3\frac{1}{2}$ in. long and $1\frac{3}{4}$ in. wide, inside.

The recesses in the end of its fork are each $\frac{3}{4}$ in. deep, and $\frac{7}{16}$ in. wide. These recesses fit over the pin that projects on each side of the stock of the crossbow, as seen in fig. 156, p. 217.

The round boxwood handle including its metal collar, is 6 in. long.

Length of the short arm of the lever including its fork, $10\frac{1}{2}$ in.   This short arm is $\frac{5}{8}$ in. wide, and $\frac{1}{4}$ in. thick.   Its fork is $3\frac{3}{8}$ in. long and $1\frac{5}{8}$ in. wide, inside. It has a half-circular notch at the end of each of its prongs.   These notches fit against the bow-string when the lever is applied to bend the bow, fig. 156.

The short arm is hinged to the long arm.   It pivots between the jaws of the block E (A, fig. 154, previous page, and fig. 155), which block is part of the long arm.

E

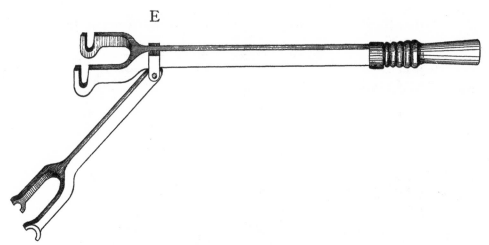

FIG. 155.—THE LEVER EXTENDED READY FOR USE.

HOW TO WORK THE LEVER, FIG. 156, OPPOSITE PAGE

1. Rest the butt of the crossbow on the ground a little in front of the right foot, the stock being in an upright position.

2. Grasp the fore-end of the stock of the crossbow, near the fore-sight, with the left hand.

3. Hook the recesses in the ends of the fork of the long arm of the lever over the $\frac{3}{8}$ in. metal pin which projects about $\frac{1}{2}$ in. on each side of the stock of the crossbow, in front of the bow, fig. 156.

4. Place the half-circular notches at the ends of the fork of the short arm of the lever against the centre of the bow-string.

5. As you hold the fore-end of the stock tight in the left hand, push with your right hand the handle of the long arm of the lever downwards towards the ground.

This action will cause the short arm of the lever to force the bow-string along the groove of the stock (fig. 156), till it finally meets and is safely secured by the catch of the lock.

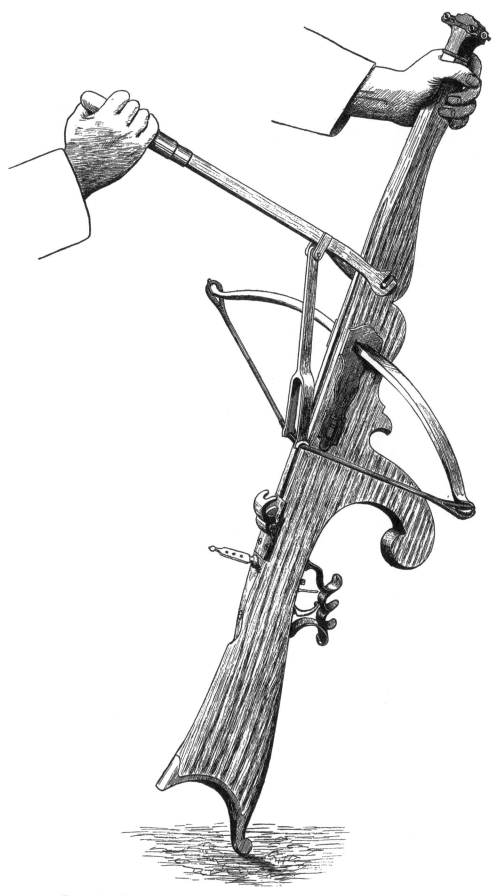

FIG. 156.—BENDING THE BOW OF THE BELGIAN TARGET CROSSBOW.

When the bowstring is stretched, and secured by the catch of the lock, the lever has, of course, no strain upon it and may be removed till again required. The cross-bow is then ready for use.

It takes but a few seconds to fit the lever to the crossbow and bend its bow.

FIG. 157.—THE TARGET-SHOOTING BOLT FOR THE BELGIAN CROSSBOW.
Half full size. Total weight, 1 oz. and 6 drachms, avoirdupois. Weight of metal head with its collar = 3s. 9d. in silver coin of the realm.

NOTE.—The bolt used with this crossbow for shooting at small wooden birds (fixed on a high pole), has a blunt head, so that when a bird is struck it is possible to knock it off its perch on the pole. See fig. 165, p. 227.

If a sharp-headed bolt were employed, it might remain in the bird without bringing it to the ground.

CHAPTER XLVI

THE BULLET-SHOOTING TARGET CROSSBOW WITH A BARREL— AS NOW USED IN BELGIUM

FIG. 159, next page, represents one of these powerful and highly finished crossbows and explains its general mechanism.

It is an adaptation of the sixteenth century slurbow shown in Chapter XXIX, which discharged a bolt.

This weapon is, however, constructed for bullets, which it shoots with great accuracy and considerable force up to fifty yards.

The bore of its steel barrel is $\frac{5}{8}$ in. in diameter. Its spherical bullet is equal in weight to 3*s*. 9*d*. in silver coin of the realm.

The steel bow is 4 in. longer—and $\frac{1}{4}$ in. wider and $\frac{1}{8}$ in. thicker at its centre—than the bow of the crossbow last described, its power being proportionately greater.

It will be seen that the barrel has a slot cut for some distance through its centre to allow the bow-string to travel to or fro.

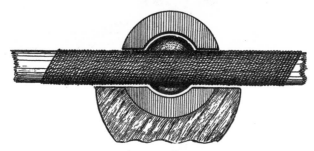

FIG. 158.—SECTION OF THE BARREL OF A BULLET-SHOOTING BELGIAN CROSSBOW.
The bullet and part of the bow-string are also shown.[1]

When the bullet is in the barrel upon the stretched bow-string, its upper and its under surface each fit, for a depth of about $\frac{1}{8}$ in., into the grooved halves of the barrel. The impact of the released bow-string comes, therefore, against the centre of the bullet when the crossbow is discharged, fig. 158.

[1] The barrel and the bullet are here given larger than full size so as to indicate the position of the latter clearly. The bullet is shown distinct from the barrel for the same reason. The bullet should fit exactly, but not so closely that it will not roll down to the bow-string when it is inserted in the muzzle of the barrel.

FIG. 159.—BULLET CROSSBOW WITH A BARREL.

The lock, and the catch that holds the stretched bow-string are the same as those shown in Chapter XLIII.

The fingers of the catch are recessed into the barrel sufficiently to allow them to move up or down as required, so that they may grasp or set free the bow-string, fig. 159.

The bow of this crossbow is bent in the same manner as the one depicted in fig. 156, p. 217. The stringing lever is also similar in all respects.

When the bow has been bent by the lever the bullet is inserted in the barrel, down which it rolls till it rests against the centre of the bow-string, fig. 158.

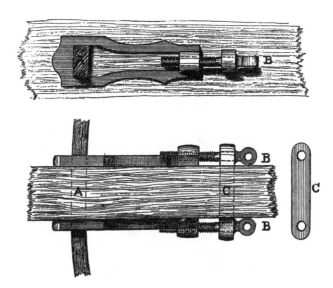

FIG. 160.—METHOD OF FASTENING THE BOW TO THE STOCK BY A METAL SCREW-STRAP.

A, the bow ; B, B, screws ; C, the crosspiece through the stock of the crossbow in which
the smooth parts of the shanks of the screws revolve.

The crossbow cannot be aimed downwards, or the bullet will run out of the barrel.

This bullet crossbow with a barrel is very popular in Belgium for shooting at small wooden birds set on the top of a pole about 100 ft. in height.

The competitors endeavour to knock the dummy birds off the pole, the shooter who succeeds in bringing down the last remaining bird winning the first prize. Unless the aiming is very correct, a bird may be struck many times without being brought to the ground.[1]

[1] The extreme top of the pole has a crosspiece. On this crosspiece there are fixed at intervals, a dozen or so sets of small outward-curving steel springs, like the feathers of a shuttlecock pointed upwards. A dummy bird, made of lignum vitæ, is placed inside each set of springs and cannot well be knocked out of

This crossbow shoots with more force than the English bullet crossbow, the reason being that it has a single bow-string which acts directly on the projectile. The Belgian weapon requires, however, a separate lever to bend its bow, and besides this inconvenience it is heavy in comparison with the one of English make which discharges a bullet.

As regards accuracy for rook-shooting, the Belgian crossbow is quite equal to a rifle, and its bullet will knock a rook lifeless without cutting it to pieces.

The extreme distance this crossbow throws its bullet, is about 380 yards.

the projections that surround it unless struck fair in the centre of the breast. The birds have different values according to their position. The pole, by means of a heavy weight attached to its butt-end, is balanced on a cross-pin which perforates it about 20 ft. from the ground, the pin being fixed between two perpendicular posts supported by props. A rope secured to the top of the pole, enables it to be swung down parallel with the ground, so that the birds may be replaced when they have all been knocked off. The pole acts like the mast of a barge when it is lowered to pass under a canal bridge.

The pendant shown above forms part of a richly decorated silver collar made by Johan Stoffel in 1600. The collar was presented to the company of crossbowmen of Eukhuizen by François Maelson, burgomaster of the town, councillor to Prince Maurice and Ambassador in England.
On the reverse side of the figure of the popinjay is engraved,

'As God ordains so it happens.'

CHAPTER XLVII

THE POPINJAY, WITH NOTES ON THE ANCIENT COMPANIES OF CONTINENTAL CROSSBOWMEN

NUMEROUS companies of citizen crossbowmen were formed in France and Belgium during the fourteenth and fifteenth centuries for the protection of their individual towns.

FIG. 161.—THE COMPANY OF ST. GEORGE.

After a fresco in an ancient Chapel of St. John and St. Paul at Ghent. From L.-A. Delaunay.

These companies of skilled and often knightly crossbowmen, especially the Company of St. George, were granted many rights and privileges and even landed estates.[1]

Delaunay writes :[2]

'Documents of the fifteenth century, prove that companies of crossbowmen were established at that period in almost all the chief towns of the provostships of Lorraine and the Barrois.

'At Liège the crossbowmen were divided into two companies, consisting of the Young and the Old Crossbowmen. The latter company was distinct from the former, its patrons being the Virgin and St. Lambert. The Young Crossbowmen had St. Hubert for their patron. The company of the Old Crossbowmen was suppressed in 1467, and its charters annulled by the Duke of Burgundy who had just subdued the country.

'This company was re-organised in 1482, but was disbanded by the Prince Bishop Ferdinand of Bavaria who confiscated its property and estates.

FIG. 162.—MEDAL OF THE GRAND ASSOCIATION OF CROSSBOWMEN OF BRUSSELS. 1560.
From L.-A. Delaunay.

'It was restored to its ancient functions and privileges in 1676, but was finally dissolved in 1684 by the Prince Bishop Maximilian of Bavaria.'

Ever since the beginning of the sixteenth century when the crossbow was supplanted by the hand-gun in warfare, the former weapon has been popular in parts of the Continent for shooting at a mark.

The mark at which the crossbow was chiefly used was the popinjay. In Belgium small figures of birds are still set up on high poles for crossbowmen to aim at. Footnote, pp. 221, 222.

The English word 'popinjay' is a corruption of 'papegai,' the old French name for a parrot.

---

[1] The members of the companies of crossbowmen were held superior in rank to ordinary troops, as they were employed to guard the person of the Sovereign in peace and war, and at all State pageants.

[2] L.-A. Delaunay. *Etude sur les Anciennes Compagnies d'Archers, d'Arbalétriers et d'Arquebusiers.* Paris, 1879. This fine and exhaustive work contains the history of all the well-known companies of crossbowmen and archers that formerly existed in France and Belgium. It is beautifully and profusely illustrated.

In confirmation of this, it is worth notice that the modern popinjay of Belgium is usually painted green, and archery records show that it has nearly always been coloured in this way.

Shooting with a crossbow at the popinjay is, perhaps, one of the oldest sports in existence in which the bow is concerned.

FIG. 163.—SHOOTING AT THE POPINJAY.

From an Illustrated Manuscript of about 1320 *in the British Museum, reproduced by J. Strutt in ' Sports and Pastimes of the People of England,'* 1801.

We even read of shooting at the popinjay in the thirteenth century, and from that time to the present day it has been a common amusement in parts of the Continent.

In Virgil's fifth book of the Æneid we find a description of shooting at a popinjay—in this case a live bird.[1]   Virgil tells us that Æneas, when voyaging to Italy, was forced by a tempest to anchor at Drepanum, a harbour on the shores of Sicily.   Here he celebrated the anniversary of the death of his father Anchises, and arranged upon the occasion, as was usual, funeral games in honour of his memory.

The games consisted of competitions in foot-racing, boat-racing, boxing and archery.   In the archery contest, the final one of the celebration, the competitors discharged their arrows at a bird fastened to the top of a mast.

> ' This done, Æneas orders for the close
> The strife of Archers with contending bows.
> The mast, Sergestus' shattered galley bore,
> With his own hands he raises on the shore.
> A fluttering dove upon the top they tie,
> The living mark at which their arrows fly.'—DRYDEN.

[1] Virgil—Roman poet, born B.C. 70, died B.C. 19.

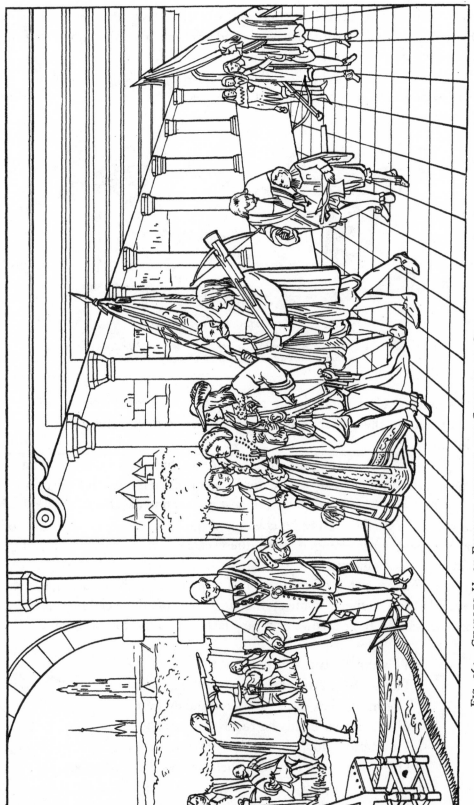

FIG. 164.—CHARLES II OF ENGLAND VISITING THE COMPANY OF CROSSBOWMEN OF ST. GEORGE AT BRUGES.

Painted by Eugène Legendre, and now at Bruges.

From L.-A. Delaunay.

It was formerly the custom in France and Belgium (as it now is in Saxony Chapter XLVIII), for each company of crossbowmen to hold an annual *fête*, and decide thereat who of their members should be 'King of the Crossbowmen,' *i.e.* King of the Bird, for the ensuing year.

The 'King' was the best marksman of a company or the one who succeeded in winning the first prize of the meeting.

FIG. 165.—PORTRAIT OF A 'KING OF THE BIRD' OF THE COMPANY OF ST. GEORGE AT BRUSSELS. (Seventeenth Century.)

From L-A. Delaunay.

These *fêtes* were carried out with much pomp and formality, and with their accessories of uniforms, banners and music were very picturesque.

As it was always an advantage to the crossbowmen to have a royal or noble patron for their 'King,' this matter was nicely arranged. If it was desired to obtain some Royal personage as 'King of the Crossbowmen' of a company, and if he happened to be a poor marksman or, if a good one, unable to attend the *fête*, then the best shot in the company acted as his deputy, and by shooting till he struck down the popinjay earned the coveted patronage.

Delaunay, in his splendid work on 'The Ancient Companies of Crossbowmen, Archers and Arquebusiers' tells us that kings and princes frequently took part in contests with the crossbow and in shooting at the popinjay.

Charles VII. used to play chess and shoot with the crossbow. Philip the Good attended the contests to encourage the knights by his presence.

The latter may be seen thus portrayed in the galleries of Lille, Bruges, Dijon, Beaune and Châlon-sur-Saône.

'Henry II., Charles IX. and Henry IV. were also present at the various performances of the companies of crossbowmen. Du Guesclin in his youth, on the field of Tacquet at Rennes, won the prize in shooting at the popinjay.

FIG. 166.—MEDAL CAST TO COMMEMORATE THE SUCCESS OF THE INFANTA ISABELLA WITH THE CROSSBOW IN 1615.

From L.-A. Delaunay.

'At twelve years of age, Charles V. brought down the popinjay in a shooting match at the Grand Serment in Brussels, and was proclaimed " King of the Crossbowmen." Charles II. of England, during his exile at Bruges, was also " King of the Crossbowmen."

'At Dijon, in June 1595, according to the Burgundian historian Courtépée, Henry IV. found delight in shooting at the popinjay.'

I add another extract from Delaunay which gives a very interesting account of shooting at the popinjay with a crossbow.

'The Infanta Isabella,—who was Governor of the Low Countries from 1598 to 1633,—was proclaimed in 1615, 'Queen of the Crossbowmen' of the Grand Lodge of Brussels. The following, according to the account of Gerard Van Loon, were the circumstances under which this took place. In May 1615, the Grand Confraternity of Crossbowmen of Brussels made great preparations for a match with the crossbow at a large leather popinjay, according to custom, at the cemetery of Notre-Dame du Sablon. The Archduke Albert and the Infanta Isabella his wife had been invited to the sports. We know from the chronicles of Brussels that Isabella readily took part in the recreations of the people. The 15th of May, the day fixed for the celebration, all eyes were turned towards the Archduchess, who, standing by the side of her husband in the midst of the crowd of crossbowmen, took the bended crossbow and after sighting for a short time let the arrow fly. Whether by luck or by skill, to the inexpressible delight of all present, she brought down the bird though it was set up as high as the steeple. A universal shout of joy rose to heaven more quickly than that happy arrow fell back again to earth. It seemed as if every one imagined that he had himself struck the bird through the hands of his sovereign, who, in the midst of all that applause and without losing any-thing of her usual dignity, accepted the Kingship of the Confraternity and did not disdain to become a simple citizen amongst simple citizens. The Princess was conducted in triumph to the high altar of the Sablon Church and decorated with the insignia of her new dignity.

'And to make it plain that she adopted, so to speak, the company of crossbowmen, she gave every member of the Confraternity a robe of silk of her own colours richly laced with gold. She also had built for them close to her palace, a magnificent club house, that she might the more conveniently attend their assemblies in her quality of Queen and direct their *fêtes* and feasts.

'The memory of this singular event has been preserved by a medal. [Fig. 166.]

'On the obverse, the bust of the Infanta covered with the richest ornaments.

'On the reverse, her monogram joined with her husband's between two crossbows, below this a Saint George the patron saint of the Confraternity. The monogram is surmounted by a crown, with the figure of the popinjay she brought down. The initial letters of the monogram represent the names Albert and Isabella, with the date 1615.

'The great success obtained by the Infanta Isabella on the 15th of May, 1615, is the subject of a picture by Antony Sallaert, a Flemish painter who was born in 1590 and lived later than 1648. This picture is preserved in the Museum at Brussels.

'The municipality of Brussels on the occasion of this remarkable event, made the Archduchess a present of twenty-five thousand florins, which she graciously accepted with every mark of gratitude. Isabella proposed that the income arising from this sum should serve to provide annually a dowry for six marriageable young girls, or to facilitate their entry into a convent if their inclination lay towards life in religion. It was then that a procession was institued in which the "Virgins of the Sablon," as they were called, were to appear, conducted solemnly by the clergy and accompanied by the Lodges. The Archduke took part in the first procession, which was arranged in accordance with the Archduchess's decision. This ceremony is depicted in a second painting by Sallaert.'

FIG. 167.—THE STONE CAPITAL OF A PILLAR IN THE CHURCH OF ST. SERNIN
AT TOULOUSE

From L-A. Delaunay.

CHAPTER XLVIII

THE CROSSBOWMEN OF DRESDEN—PRIVILEGIRTE
BOGENSCHÜTZEN-GESELLSCHAFT

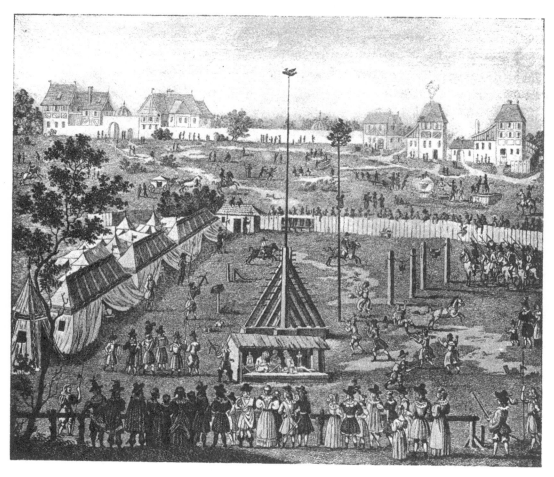

FIG. 168.—CROSSBOW-SHOOTING AT DRESDEN IN 1612.[1]

CROSSBOW-SHOOTING is as popular in Dresden and in a few other parts of Saxony, as it is in Belgium and, to a lesser degree, in the north of France.

The crossbow societies of Dresden were established in mediæval days and one of them is still in a flourishing condition.

[1] From an engraving by J. Kellerthaler, entitled 'Dresdner Vogelwiese im Jahre 1612. Kellerthaler was born about 1550. The Vogelwiese was the bird-field in which crossbow-shooting at the large wooden bird (as shown above) took place.

There are two societies of crossbowmen in Dresden, both of considerable antiquity. The Bogenschützen-Gesellschaft [1]—the older and more important of the two with a membership of nearly four hundred—is styled 'privilegirte,' which implies Royal patronage.

As this Society has ancient privileges and interesting associations, I will briefly narrate its history and describe the manner of shooting its members now practise with the crossbow.

The records of the Society date from 1416, and it is said to have been founded in 1286, the date its club flag displays.

The ' Bogenschützen-Gesellschaft ' of Dresden originally consisted of a kind of militia or municipal guard of crossbowmen, which was raised for the defence of the town in case of attack by an enemy.

From days immemorial, the Society has been endowed with corporate rights and other privileges by the Sovereign as well as by the town council. When the crossbow was laid aside, as no longer of use in warfare, the Sovereign and town council, however, continued their patronage of the Society and took part in its annual *fêtes*.

The crossbowmen were shown special favour, were supported by grants of land and money, and were employed at all State ceremonies and popular festivals, on which occasions and especially in processions they were under the control of the town council.

The Diet of the Kingdom of Saxony recognised the corporate rights of the Society, and this recognition even found expression in legislation.

At the present time, it is curious to note, the rules of the Society are subject to the approval of the Saxon Government.

In recent years the Town Council of Dresden, under strong protest from the crossbowmen, have discontinued their ancient subsidies, and withdrawn from that control of the Society they previously exercised for several centuries.

In consequence of these alterations, its members were obliged to purchase a new plot of land for their meetings.

This consists of a large field near the Elbe, outside Dresden, which has been used for the annual competitions with the crossbow for the past twenty-seven years.

--- --- ---

The Society owns several valuable medals. These were struck from time to time to celebrate notable events in its history that have occurred since

[1] Company, society, or guild of archers.

its formation.   For instance, in July 1676 the Crown Prince of Saxony made the 'Königschuss,' or King's shot at the Bird, a feat which entitled him to be 'King of the Crossbowmen' for the year.

To commemorate this incident, a large gold medal worth forty-six ducats was struck.   This is annually worn at the banquet of the Society by the member who made the 'Königschuss' at the last 'Vogelschiessen.'

The tall pole with the bird on its summit is shown on the reverse side of the medal.

In 1707, the English Ambassador in Dresden made the 'Königschuss,' with the result that Queen Anne of England ordered a medal to be coined of the value of twenty ducats, as a memento of the incident.

It has long been the custom for the King of Saxony—attended by his Court and by any members of the Royal Family who happen to be in Dresden—to discharge the opening shot of the annual competition.

The late King of Saxony, either personally or by deputy, frequently shot off the first crossbow bolt at the bird on the pole.

NOTES ON THE REGULATIONS AND PRIVILEGES OF THE SOCIETY OF CROSS-BOWMEN OF DRESDEN ; 'BOGENSCHÜTZEN-GESELLSCHAFT'

Election is by ballot, the candidate being introduced by a member.

All men of decent education, respectable antecedents and independent means are eligible.

The affairs of the Society are administered by two presidents and a committee of six members, fees ranging from 2*l.* to 10*l.* being paid to these officials for their services.

Members of the Saxon Royal Family are elected from the day of their birth and annually pay certain fees to the Society.

In addition to the yearly *fête*, or 'Vogelschiessen,' the Society gives a banquet and a ball.

In the absence of one of the Royal Family, the member who made the 'Königschuss,' or King's shot, of the current year occupies the seat of honour at the banquet and receives the guests.

The Court of Saxony and the municipality do not now pay their annual subsidies to the Crossbowmen as they did formerly.

The municipality, however, under ancient contracts pays a considerable sum per annum as compensation for certain building and repairing charges at one time undertaken by the Society.

The original grant of wine to the crossbowmen from the Royal cellars, for use at their *fêtes*, has been exchanged for a sum of money, out of the interest of which gold and silver medals are provided as prizes.

Wine from the Royal cellars is still supplied to the members for their banquet and ball.

The 'Bogenschützen' competitions take place every year and are held in the week falling at the end of July or beginning of August, the meeting commencing on a Monday and terminating on or before the following Saturday.

THE BIRD AT WHICH THE CROSSBOWMEN SHOOT

The target at which the members of the Society shoot with their crossbows consists of a large and gaily coloured figure of a bird, made of wood and somewhat resembling the Imperial Eagle of Germany.

The bird is 13 ft. in length from head to tail and 8 ft. in breadth across its extended wings. Its weight is 200 lb. It is fixed to the top of a mast 136 ft. in height.

The mast is laid on the ground and the bird is bolted to its smaller end. The mast is then raised into an upright position by a large number of men pulling the ropes attached to it. To secure the mast from falling, it is fixed into an immense framework in the form of a trestle, fig. 114, p. 176. It is also further secured by ropes fastened to posts driven into the ground.

The bird is composed of numerous pieces of various shapes and dimensions.

About fifty of these pieces have distinctive names and the crossbowman who brings one down is paid in accordance with its value as a prize.

For example he receives 6s. for the sceptre or the orb, a medal for the silvered ball in the crown, 4s. for the beak, 12s. for a wing joint and other sums for different feathers.

A gold medal is given to the member who detaches the top feather on the right-hand or left-hand side of the tail.

Pieces of thick glass ('Kleinode' or 'jewels') are inserted in various parts of the bird, three of which are in the tail and one in each wing.

Each 'jewel' has its name engraved on its reverse side so that it can easily be assigned to the marksman who knocks it off.[1]

---

[1] Besides the more valuable prizes (which are given for the special parts of the bird), a small sum of money is paid for every fragment of the bird that is knocked off. All such pieces are paid for in proportion to their weight, none being recognised that weigh less than 20 grammes.

The crossbow bolts break the entire bird to fragments by degrees and bring it down in larger or smaller pieces, the heart or centre being always the last part to fall.   A new bird has, of course, to be made every year for the annual competition.

The crossbowman who shoots off the last remaining portion of the heart or centre of the bird and who thus makes the 'Königschuss' or King's shot, is the champion for the year.   He receives 7*l.* 10*s.* in cash, out of which he has to pay a number of *douceurs* to the attendants employed.

The competitors stand at a distance of 50 yards from the foot of the mast, and each member shoots in regular rotation one bolt at a time.

From the foregoing account it will be seen that many hundred bolts may be expended before the bird is completely destroyed.   The meeting is intended to last a week and it usually does so.

The crossbow used in the competitions at Dresden, as well as for frequent practice throughout the summer at ordinary circular targets, is, together with its lever, the same in shape and mechanism as the one shown in fig. 105, p. 165.

The Dresden crossbow is, however, larger, and has a considerably stronger steel bow

FIG. 169.—THE DRESDEN BIRD.

than that given in fig. 105 ; the more modern ones having also butt-ends like fowling-pieces.   The Dresden weapon costs from 5*l.* to as much as 15*l.*

The bolt—'Kronenbolzen'—employed with the Dresden crossbow for shooting at the bird has a heavy and blunt metal head.   The head, as the name implies, is in the form of a crown, with four outside points and a centre one all on the same level,[1] fig. 170.

---

[1] If the bolt were sharp at its end, like an arrow, it would stick into the parts of the bird and hence fail to knock them off.   As the bolts fall to the ground they are rapidly passed back to the shooters by a number of children, who, standing in line at short intervals, adroitly throw them from hand to hand till they finally reach their respective owners.   This method of returning the bolts is even indicated in fig. 168, p. 231.

The great Dresden fair, of which the crossbow competition forms a part, annually attracts many thousands of visitors. It is the week of the year for the people of the towns and country of Saxony, and from the amount of feasting and gaiety it entails, is sometimes called the 'tolle Woche' or mad week.[1]

The old custom of electing as their King the most successful marksman of a company of crossbowmen, longbowmen or arquebusiers, prevailed in many parts of the Continent of Europe besides France, Belgium and Saxony.

For instance, John Evelyn during his visit to Geneva in 1646, writes:

'A little out of the Towne is a spaceous field which they call Campus Martius, . . . for here on every Sonday after the evening devotions, this precise people permitt their youths to exercise armes and shoote in gunns and in the long and crossebowes, in which they are exceedingly expert, reputed to be as dexterous as any people in the world. To encourage this they yearely elect him who has won most prizes at the mark to be their king, as the king of the long-bow, gun or crossebow. He then wears that weapon in his hat in gold with a crowne over it, made fast to the hat like a broach. In this field is a long house wherein their armes and furniture are kept in severall places very neately. To this joynes a hall where at certain times they meete and feast; in the glass windows are the armes and names of their kings of armes.'[2]

---

[1] For many of these notes on the crossbowmen of Saxony I am indebted to Sir Condie Stephen late Resident British Minister at Dresden, to Mr. H. J. Stanley recently vice-consul and to Hofrath Dr. Peschell of the Körner Museum, Dresden.

[2] *Diary of John Evelyn*, edited by Henry B. Wheatley, F.S.A., i. 290-291.

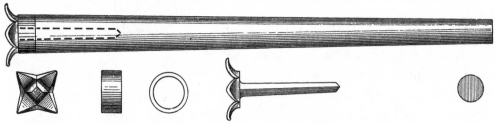

FIG. 170.—DRESDEN CROSSBOW BOLT (KRONENBOLZEN). Half full size.

Total weight, 2¾ oz. Weight of metal head and collar, 1¼ oz. The balancing-point of the bolt is 2½ in. from its head-end. Though the bolt has no feathers it flies accurately from the crossbow.

CHAPTER XLIX

THE CHINESE REPEATING CROSSBOW

HERE we have surely the most curious of all the weapons I have described.

Though the antiquity of the repeating crossbow is so great that the date of its introduction is beyond conjecture, it is to this day carried by Chinese soldiers in the more remote districts of their empire.

In the recent war between China and Japan, 1894–95, the repeating crossbow was frequently seen among troops who came from the interior of the first-named country.

The interesting and unique feature of this crossbow is its repeating action, which though so crudely simple acts perfectly and enables the crossbowman to discharge ten arrows in fifteen seconds.

When bows, and crossbows which shot one bolt at a time, were the usual missive weapons of the Chinese, it is probable that the repeating crossbow was very effective for stopping the rush of an enemy in the open, or for defending fortified positions.

For example, one hundred men with repeating crossbows could send a thousand arrows into their opponents' ranks in a quarter of a minute.

On the other hand, one hundred men with bows, or with ordinary crossbows that shot only one arrow at a discharge, would not be able to loose more than about two hundred arrows in fifteen seconds.

The effect of a continuous stream of a thousand arrows flying into a crowd of assailants—in so short a space as fifteen seconds—would, of course, be infinitely greater than that of only two hundred in the same time, especially as the arrows of barbaric nations were often smeared with poison.

The small and light arrow of the comparatively weak Chinese crossbow here described had little penetrative power. For this reason the head of the arrow was sometimes dipped in poison, in order that a slight wound might prove fatal.

The impetus of the heavy bolt of the mediæval European crossbow which had a thick steel bow, was sufficient to destroy life without the aid of such a cruel accessory as poison.

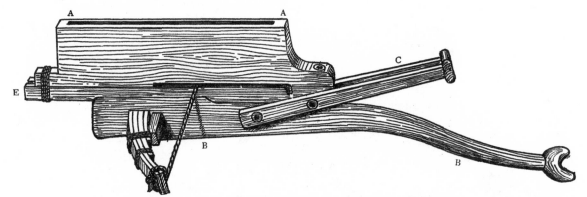

FIG. 171.—SIDE VIEW OF THE CHINESE REPEATING CROSSBOW.

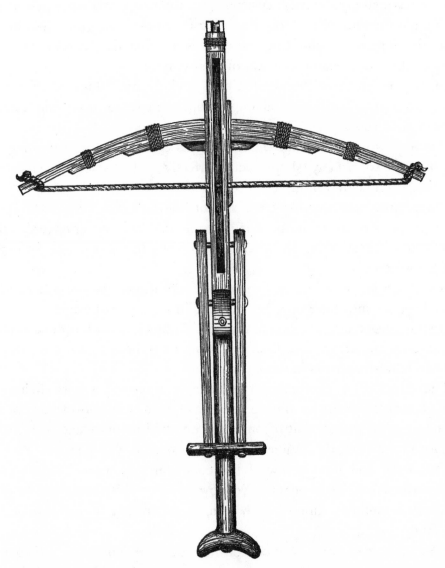

FIG. 172.—SURFACE VIEW OF THE CHINESE REPEATING CROSSBOW, SHOWING THE
OPENING AT THE TOP OF ITS MAGAZINE.

THE CONSTRUCTION OF THE CHINESE REPEATING CROSSBOW, FIGS. 171, 172

A, A. The magazine in which the ten or twelve small arrows are laid (one on the other) when the weapon is made ready for use.

B, B. The stock in which the bamboo bow is fixed.

C. The lever that works the crossbow. The lever is hinged to the stock of the crossbow and its magazine by metal pins, fig. 174, next page.

E. The piece of wood along the upper surface of which a groove is cut for an arrow to rest in, and that also has a notch in it to hold the bow-string.

This piece is attached to the magazine and forms the lower part of it.

HOW TO WORK THE CROSSBOW, FIG. 174, NEXT PAGE

By pushing forward the magazine by means of the lever, the bow-string is automatically caught in the notch above the trigger, A, fig. 174, next page.

At the moment when the bow-string is thus secured, an arrow falls from the magazine into the groove cut out in front of the notch. An arrow cannot drop from the magazine into the groove till the bow-string is in the notch, fig. 175, p. 242.

The trigger consists of a little piece of hard wood. When the lever is fully pulled back the trigger pushes the stretched bow-string upwards out of the notch that holds it, B, fig. 174, next page. The trigger works in an upright slot. It has its upper end enlarged to prevent it from dropping out of the slot in which it moves up or down, fig. 173.

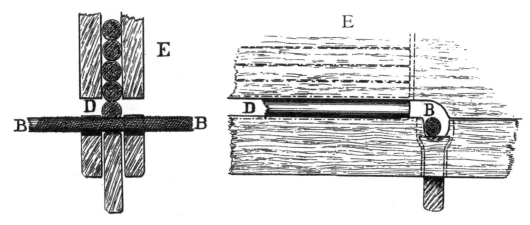

FIG. 173.—THE ACTION OF THE TRIGGER OF THE CHINESE REPEATING CROSSBOW.

B, The bow-string in the notch above the trigger ; D, An arrow in the groove in front of the bow-string ; E, The magazine which contains the supply of arrows.

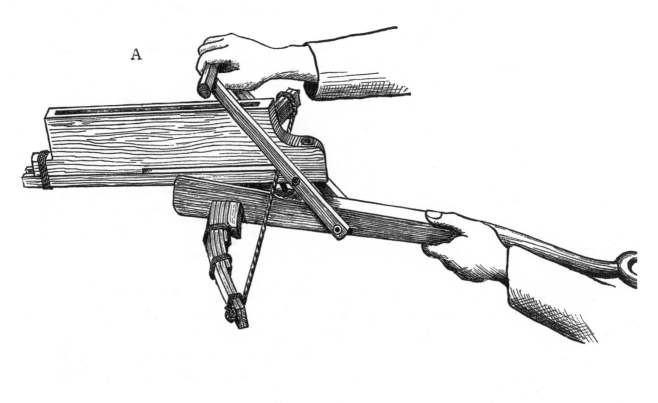

A

B

FIG. 174.—THE ACTION OF THE CHINESE REPEATING CROSSBOW.

A. The magazine, full of arrows, pushed forward by the lever.   The bow-string is caught in the notch above
   the trigger.
B. The crossbow just before it is discharged.   The trigger, as its lower extremity is pressed against the
   surface of the stock by the action of the lever lifts the bow-string out of the notch.

B, fig. 174. The lever is here pulled back, with the result that the bow is bent and the bow-string stretched. By pulling back the lever a little farther than shown in this sketch, the projecting end of the trigger will be pressed against the surface of the stock of the crossbow. This causes the upper end of the trigger to lift the bow-string out of the notch and set it free. The arrow is then discharged and the crossbow returns to the position shown in fig. 171, p. 238, and is ready for the next shot.

From this description, it will be understood how simple and rapid is the action of the crossbow. All that need be done to shoot off the arrows contained in its magazine, is to work the lever to and fro as slowly or as quickly as desired.

It is even possible to discharge a dozen arrows in fifteen seconds.

By a slight alteration in the construction of the crossbow it was sometimes made to shoot two arrows, instead of one, every time its bow recoiled.

In such a case, the magazine and stock were about $\frac{3}{4}$ in. wider than in the weapon just described. The magazine had a thin partition down its centre which divided it into two compartments. On each side of the central partition a dozen arrows were laid, one over the other. The bow-string passed over two parallel grooves instead of over a single one, each groove being, of course, exactly beneath a compartment in the magazine. As the lever was worked, two arrows dropped from the magazine and remained side by side, one in each groove, both arrows being propelled together when the bow-string was released.

By means of this arrangement one hundred men could discharge two thousand arrows in fifteen seconds, or double the number which one hundred men could shoot off in the same time with the ordinary repeating crossbow.

The effective range of these Chinese weapons was about 80 yards; their extreme range from 180 to 200 yards. The bamboo arrows, though short and light, were well made and had steel heads that were heavy in proportion to the length of their shafts. They had no feathers, so that their freedom of movement might not be impeded as they dropped one by one from the magazine when the crossbow was being used.

For the same reason, the width of the magazine—inside—was slightly in excess of the diameter of the arrow.

The length of the arrow was from 12 in. to 16 in., according to the size of the crossbow; its diameter $\frac{5}{16}$ in. to $\frac{3}{8}$ in.

The bow was made either of one stout piece of male bamboo, about 3 ft.
6 in. long, or of several flat strips lashed together.

In the latter case, the bow-string passed through a hole in each end of the
bow, fig. 174, p. 240.   The bow-string consisted of animal sinew twisted into a
cord of suitable strength.

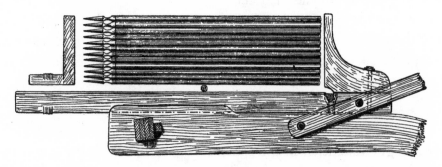

FIG. 175.—THE MAGAZINE OF THE CHINESE REPEATING CROSSBOW WITH ITS SIDES
REMOVED.

It will be seen that an arrow cannot drop down from the magazine into the groove along which
the bow-string travels till the latter is in the notch above the trigger, as shown in A, fig. 174,
p. 240.

CHAPTER L

ARROW-THROWING

In connection with long-distance shooting with the bow or the crossbow, it is interesting to consider to what a surprising range an arrow can be thrown by hand, with the mere assistance of a little piece of string.

In a few parts of the West Riding of Yorkshire, the ancient pastime of arrow-throwing is a popular sport among the pitmen. Sometimes as many as a couple of thousand spectators witness a contest between two noted performers.[1]

The matches are decided by the aggregate distance of an equal number of throws (usually from twenty to thirty) by each competitor.

If a man sends his arrow let us say, 12 score, 13 score, 14 score yards, in three throws, then his total is 39 points.

The ordinary thrower will cast the arrow from 240 to 250 yards, a very skilful thrower will send it from 280 to 300 yards, the record throw being 372 yards.

A short time ago I entertained a party of arrow-throwers. Several of them achieved a range of from 270 to 280 yards, as I carefully verified at the time with a surveyor's chain.

The arrow is shaped from a rod of straight hazel wood of about the thickness of a man's little finger, the hazel rods with dark-coloured bark being preferred for making the arrows.

The rods are laid aside in a dry place for two years before they are shaped into arrows. Though the finished arrow has neither metal head nor feathers nor any form of nock, yet it flies through the air with the true and graceful curve of the best arrow used in archery.

---

[1] An arrow-thrower will often practise daily for several weeks before he engages in a match, in order to accustom his arm to stand the strain required of it when throwing the arrow.

It is also necessary that an arrow-thrower should train his eye to give him the correct angle at which to cast the arrow, so that he may obtain as long a range with it as his strength of arm admits.

In these arrow-throwing matches a considerable amount of money is lost or won, and the couple of men competing often throw for a stake of from £20 to £30, that has been subscribed for them by their supporters.

The hazel arrow is slightly tapered from one end to the other, its bluntly pointed thick extremity forming the head, or part which travels forward when the arrow is thrown. The pith of the hazel forms the longitudinal centre of the arrow. The arrow is, in fact, formed of pith with a thin shell of wood outside it.

The length of the arrow is 31 in. Its diameter at its small end $\frac{3}{16}$ in. ; at its centre of length, $\frac{1}{4}$ in. ; near its head, $\frac{5}{16}$ in.; see fig. 177, p. 246.

Its weight is a little over $\frac{1}{2}$ oz., or equal to 3s. 6d. in silver coin of the realm.

The balancing point of the arrow is 13 in. from its head, A, fig. 176.

The arrow-throwers are so particular in regard to the balance of their arrows, that they may sometimes be seen inserting a common pin up the pith of the small end of an arrow, or else perhaps extracting one.

The trivial difference of the weight of an ordinary pin will aid or retard the flight of one of these hazel arrows by several yards, and may be the cause of a match being lost or gained as the case may be.

TO THROW THE ARROW, FIG. 176

1. Make a pencil mark round the arrow at 16 in. from its head, B, fig. 176.

2. Take a piece of hard strong string, $\frac{1}{16}$ in. in diameter and 28 in. long. Tie a double knot at $\frac{1}{2}$ in. from one end of the string, C, fig. 176.

3. Hold the head of the arrow towards you in your left hand, and hitch the knot firmly round the pencil mark, as shown at D, fig. 176.[1]

4. Next, and still holding the head of the arrow towards you in the left hand, twist the loose end of the string round the first joint of the first finger of the right hand, until the inside edge of this finger is 3 in. from the point of the arrow along its shaft. Keep the string meanwhile tightly stretched from the finger to the knot. The knot will not slip if the string is kept taut, E, fig. 176.[2]

5. Now grip the arrow close to its head between the thumb and the second and third fingers of the right hand; (the first finger keeping the string tight,) and turn it from you in the direction of its intended flight, F, fig. 176.

6. Hold the arrow at arm's length in front of you, then draw it back and

---

[1] Remember that the knot is merely hitched to the arrow and not tied to it.

[2] During the process of winding the string on the forefinger of the right hand, the left hand should grasp the arrow and the string tightly together a few inches below the knot, so as to prevent the latter from slipping. The part of the string (about half its length) which is wrapped round the finger may be unravelled so as not to cut the skin. The unravelled portion may be stopped by a knot from unwinding too far.

with a powerful jerk of the arm cast it forward and high as if throwing a stone, its line of flight being at an angle of 45 degrees to the ground.

Though the movement of the arm in the act of throwing should be chiefly below the top of the shoulder, yet the arrow should be projected upwards as it leaves the hand, G, fig. 176.

If the arrow is inclined 'to one side or other of the shoulder just as it is thrown, it will only travel an erratic course of a hundred yards or so.

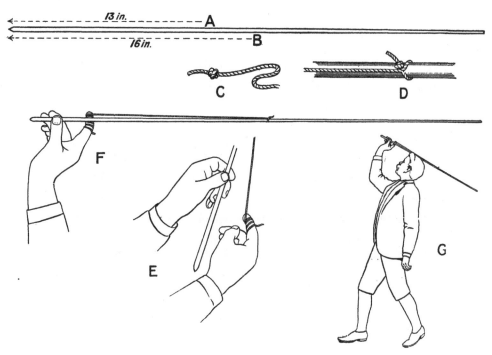

FIG. 176.—ARROW THROWING. C and D are half full size.

As the arrow flies away the knotted end of the string drops off its shaft.

The propelling power is derived entirely from the first finger, which gives the arrow its impetus by pulling sharply against the tightly stretched string as the arm is jerked forward.

The second and third fingers and the thumb merely retain the arrow in position, whilst the first finger exerts the force that propels it.

The difficulty with beginners is, to avoid gripping the arrow between the second and third fingers and the thumb at the moment when it should leave the hand. This, of course, prevents the first finger applying its force to the string, with the result that the flight of the arrow is completely checked.

I lately persuaded an arrow-thrower to show me how far he could project a light archery flight arrow by means of a piece of string.

This man had never even seen such an arrow before ; he sent it, however, several times from 180 to 200 yards.

Though I have tried all kinds of seasoned wood, light and heavy, I find no material equals dry hazel for these arrows.   A really good-flying arrow is not easy to make and a dozen may be constructed without a successful one amongst them.

The difficulty is to shape the hazel rod so that the pith inside it is longitudinally correct as regards its being in the exact centre of the finished arrow. If this is not the case the flight of the arrow will be irregular and its range a short one.   A first-class arrow is usually valued at a sovereign by an arrow-thrower.

This is easily understood when we consider that by travelling a few yards further than an arrow of inferior balance to it, a good arrow may be the means of winning a valuable prize.

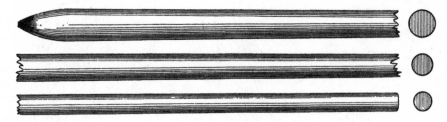

FIG. 177.—PART OF THE HEAD-END ; OF THE CENTRE ; AND OF THE BUTT-END OF A
HAZEL ARROW.   Full size.

Part IV.

A TREATISE ON THE SIEGE ENGINES USED IN ANCIENT AND MEDIÆVAL TIMES FOR DISCHARGING GREAT STONES AND ARROWS

CHAPTER LI

INTRODUCTORY NOTES ON THE SIEGE ENGINES USED IN ANCIENT AND MEDIÆVAL TIMES FOR DISCHARGING GREAT STONES AND ARROWS

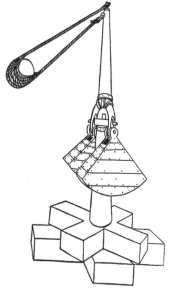

FIG. 178.—A TREBUCHET OR SLING ENGINE.

Criticism.—As the arm has been brought into an upright position by its counterpoises, the stone should have been projected from the sling. See fig. 212, p. 310.

From Valturius. Edition of 1472.

IN connection with the history of the crossbow—or as some old writers term it, the manubalista—it will be of interest to describe among other engines the balista, the original weapon from which the crossbow is said to have been adapted.

The ancient balista resembled the crossbow in its general construction, though the former —which was employed for propelling huge darts against the defenders or besiegers of a fortified castle or town—was of immensely greater size.[1]

The three projectile engines used in ancient and mediæval sieges, were the Balista, the Catapult and the Trebuchet. The first named discharged great arrows and the other two cast pieces of rock and heavy balls of stone.

The balista and the catapult date from time immemorial but the trebuchet was an early mediæval invention. All three engines were employed for many years after the first appearance of cannon in warfare.

The catapult was sometimes known as 'onager.' It was also called a 'mangonel.' The word 'scorpion' refers I consider to the balista. Procopius tells us that the catapult was termed 'onager' or 'wild ass' because it was likened to this animal, which when harried by dogs kept them off by scattering stones with its hind feet.

---

[1] The balista had not, however, a bow in one piece like a crossbow. It had two arms joined by a bowstring, each arm working independently between twisted cords. See Chapter LVII.

The balista and the catapult derived their projectile force from the recoil of tightly twisted cordage, while the trebuchet owed its power to the utilisation of the force of gravity of a heavy weight.

Though the construction of these engines was quite distinct, and the one kind projected stones and the other arrows, their respective names have been so carelessly used by many mediæval as well as later writers, that it is often impossible to tell which class of engine an author alludes to.

Even among the earliest historians, the name 'balista' was frequently bestowed on any large siege weapon that discharged missiles—whether the missiles were bolts or stones.[1]

The following names were commonly, and often indiscriminately applied to the ancient and mediæval engines that projected stones and arrows of large size :

Balista	Engin	Martinet	Scorpion
Beugle	Engin à verge	Matafunda	Springald
Blida	Espringale	Mategrifon	Tormentum
Bricole	Fronda	Petrary	Trebuchet
Calabra	Fundibulum	Robinet	Tripantum
Catapulta	Manganum		

Though so many names suggest that there were numerous varieties of siege engines, this was not the case.

All these names refer at most to four distinct weapons, and these I shall presently describe.

Besides the names given above, others were coined for certain well-known machines which from their power or accuracy became popular among the soldiery. For instance we read of the War-wolf, the Wild-cat, the Bull-slinger, the Ill-neighbour, the Queen, the Lady and so forth. Just as in our day an artilleryman will bestow particular care on the appearance of a gun in the performances of which he takes pride, so doubtless the ancients favoured one or other engine which had distinguished itself in action, and called it by some fanciful name to record its success.

Many of the illustrations of balistas and catapults to be found in late

---

[1] The catapult is often described as having been employed for throwing heavy javelins as well as great stones. In my opinion, based on practical experience, the mechanism of a catapult could not possibly have been adapted for projecting a javelin. Its construction shows that this engine was never intended for such a purpose. The name catapult was, however, often applied to the balista. This confusion was no doubt the cause of mistakes on the part of those compilers who were ignorant of the mechanical details of the two weapons. In the large picture of a Roman catapult painted by Sir E. Poynter, P.R.A., the artist has depicted a weapon that actually combines in its mechanism the parts of a catapult a trebuchet and a spring engine !

FIG. 179.—THE CAPTURE OF A FORTRESS.

Criticism.—A fortification being entered by the besiegers, who have made a breach in the outside wall with a battering ram.

A catapult is in the left corner of the picture and four men are taking a balista up the approach to the gateway.

The catapult is shown with its skein of cord between its uprights, instead of between its sides as should be the case.

The balista is also incorrect, as its arms are not in the centres of the skeins of cordage which work them.

From Polybius. Edition of 1727.

mediæval books are inaccurate through having been incorrectly copied from older works.

In some cases the machines are of absurd appearance, their details of construction having been evolved from the imagination of the artists who attempted to portray them.

In other instances authors have quoted extracts from early manuscripts, and then supplied laboriously minute engravings of what they fancied the engines referred to were like.

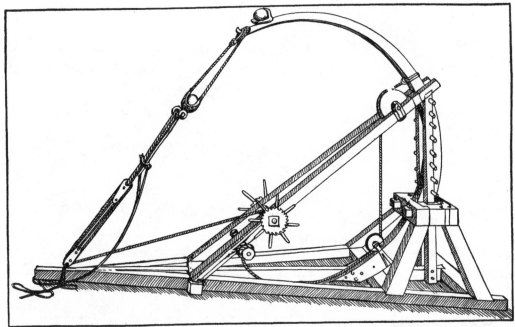

FIG. 180.—A SPRING ENGINE WITH A SLING LIKE THAT OF A TREBUCHET ATTACHED TO ITS ARM, WHICH CAST TWO STONES AT THE SAME TIME.

From 'Il Codice Atlantico,' Leonardo da Vinci. 1445-1520.

As I have made both large and small working models of the projectile siege engines of the ancients, I shall hope to elucidate their mechanism.[1] These models with their details and dimensions, are the result of a practical application of the best ancient and mediæval directions and plans I have been able to discover in home and Continental libraries.

With some knowledge of engineering and ballistics, I have not found it difficult to fit together the parts of these engines or to ascertain their mechanism and relative power. In a mediæval drawing of a balista or a catapult the perspective was commonly of ludicrous incorrectness, but by constructing a model of fair size the design of the original may often be discovered.

[1] My largest catapult weighs one and a half tons.

It frequently happens that in a mediæval picture of one of these machines some important mechanical detail is omitted, or from the difficulty of portraying it correctly is purposely concealed by figures of soldiers, an omission that may, however, be supplied by reference to other representations of the same weapon.

FIG. 181.—A TREBUCHET WITH ITS ARM WOUND DOWN BY MEANS OF LARGE HOLLOW WHEELS, WITH MEN WORKING INSIDE THEM ON THE PRINCIPLE OF THE TREADMILL.

Criticism.—The stone is in its sling on the central plank inside the framework. This plank is grooved to prevent the stone from inclining to one side or the other as it is raised by the arm of the engine.

The man with a rope in his hand is about to free the catch which releases the arm. The instant this man releases the arm, the score or so of men at the fore-part of the engine pull the ropes attached to the counterpoise, with a view to increasing the speed of its descent and hence the velocity with which the arm ascends. The commander is giving a signal with his bâton that the engine is to be discharged. I do not consider that any number of men pulling down the counterpoise with ropes would exert any influence on the speed with which so vast a weight would descend—a weight of several tons! See Chapter LVIII.

From Viollet-le-Duc.

It is indeed impossible to find a complete working plan of either a catapult, a balista or a trebuchet; a perfect design for one being obtainable only by consulting many authorities.

Some of the historians, mechanicians and artists from whom information on balistas, catapults and trebuchets may be derived, are as follows. I name them alphabetically irrespective of their periods :

ABBO : A monk of Saint-Germain des Prés, born about the middle of the ninth century, died in 923. He wrote a poem in Latin describing the siege of Paris by the Northmen in 885–886.

AMMIANUS MARCELLINUS : Military historian. Died shortly after 390. His work first printed at Rome 1474. The latest edition is that of V. Gardthausen, 1874–1875.

APPIAN : Historian. Lived at Rome during the reigns of Trajan, Hadrian and Antoninus Pius, 98–161. The best edition of his history is that of Schweighaeuser, 1785.

APOLLODORUS OF DAMASCUS : Built Trajan's Column, 105–113. Architect and engineer. Addressed a series of letters to the Emperor Trajan on siege engines (*vide* Thévenot).

ATHENÆUS : Lived in the time of Archimedes, B.C. 287–212. The author of a treatise on warlike engines (*vide* Thévenot).

BITON : Flourished about 250 B.C. Wrote a treatise on siege engines for throwing stones (*vide* Thévenot).

BLONDEL, FRANÇOIS : French engineer and architect ; born 1617 ; died 1686.

CÆSAR, JULIUS (the Dictator) : Born B.C. 100 ; died B.C. 44. Author of the ' Commentaries ' on the Gallic and Civil wars.

CAMDEN, WILLIAM : Born 1551 ; died 1623. Antiquary. Published his ' Britannia ' 1586–1607.

COLONNA, EGIDIO : Died 1316. Archbishop of Bourges 1294, after having been tutor to Philip the Fair of France. His best known works are ' Quæstiones Metaphysicales ' and ' De Regimine Principum ; ' the latter was written about 1280. Colonna gives a description of the siege engines of his time.

DANIEL, PERE GABRIEL : Historian. Born 1649 ; died 1728. Published ' Histoire de France,' 1713.

DIODORUS (The Sicilian) : Historian. Lived under Julius and Augustus Cæsar (Augustus died A.D. 14). The best modern edition is that edited by L. Dindorf, 1828.

FABRETTI, RAFFAEL : Antiquary. Born 1618 ; died 1700.

FROISSART, JEAN : French chronicler. Born about 1337 ; died 1410. His chronicles printed about 1500. Translated into English by Lord Berners, and published 1523–1525.

GROSE, FRANCIS : Military historian and antiquary. Born about 1731 ; died 1791. Published ' Military Antiquities ' 1786–1788.

HERON OF ALEXANDRIA : Mechanician. Lived B.C. 284–221. Bernardino Baldi edited his work on arrows and siege engines, 1616 (*vide* Thévenot).

ISIDORUS, BISHOP OF SEVILLE : Historian. Died 636.

JOSEPHUS, FLAVIUS: Jewish historian. Born A.D. 37; died about the year 100. Wrote the history of the Jewish wars and also Jewish antiquities. Josephus, acting as commander of the besieged, bravely defended Jotapata, A.D. 67, against the Roman general Vespasian. He was also present with the Roman army during the siege of Jerusalem by Titus, A.D. 70.

LEONARDO DA VINCI: Italian painter. Born 1445; died 1520. In the immense volume of sketches and MSS. by this famous artist, which is preserved at Milan and entitled 'Il Codice Atlantico,' there are several drawings of siege engines.

LIPSIUS, JUSTUS: Historian. Born 1547; died 1606.

MEZERAY, FRANÇOIS E. DE: French historian. Born 1610; died 1683. Published 'Histoire de France,' 1643–1651.

NAPOLEON III.: 'Etudes sur l'artillerie,' compiled by order of the Emperor and containing many drawings of the full-sized models of siege engines made by his orders, with interesting and scientific criticism of their power and effect.

PHILON OF BYZANTIUM: A writer on and inventor of warlike and other engines. Lived shortly after the time of Archimedes (Archimedes died 212 B.C.): was a contemporary of Ctesibius, who lived in the reign of Ptolemy Physcon, B.C. 170–117 (*vide* Thévenot).

PLUTARCH: Biographer and historian. Time of birth and death unknown. He was a young man, A.D. 66.

POLYBIUS: Military historian. Born about B.C. 204. His history commences B.C. 220 and concludes B.C. 146. The most interesting edition is the one translated into French by Vincent Thuillier with a commentary by de Folard, 1727–1730.

PROCOPIUS: Byzantine historian. Born about 500; died 565. The best edition is that of L. Dindorf, 1833–1838.

RAMELLI, AGOSTINO: Italian engineer. Born about 1531; died 1590. Published a work on projectile and other engines, 1588.

TACITUS, CORNELIUS: Roman historian. Born about A.D. 61.

THEVENOT, MELCHISEDECH, 1620–1692: Edited a book called 'Mathematici Veteres,' containing several treatises on the mechanics and siege operations of the ancients, including the construction and management of their projectile engines. In this book are to be found the writings on the subject of military engines that were compiled by Athenæus, Apollodorus, Biton, Heron and Philon. Thévenot was King's librarian to Louis XIV. After his death the manuscript of 'Mathematici Veteres,' or 'The Ancient Mathematicians,' was revised, and published by La Hire in 1693. The book was again edited by Boivin, an official in the King's library, who lived 1663–1726.

The treatises contained in Thévenot were finally re-edited, and published
by C. Wescher, Paris, 1869.

VALTURIUS, ROBERTUS : Military author.   Living at the end of the fifteenth
century.   His book 'De Re Militari' first printed at Verona, 1472.

VEGETIUS, FLAVIUS RENATUS : Roman military writer.   Flourished in the time
of the Emperor Valentinian II., 375–392.   The best edition is that of
Schwebel, 1767.

VIOLLET-LE-DUC : French military historian.   Published his 'Dictionnaire
raisonné de l'Architecture,' 1861.

VITRUVIUS POLLIO : Architect and military engineer, and inspector of military
engines under the Emperor Augustus.   Born between B.C. 85 and 75.   His
tenth book treats of siege engines.   Translated into French with commentary
by Perrault, 1673.   The most interesting editions of Vitruvius are those
containing the commentary on siege engines by Philander.   The best of
these is dated 1649.

FIG. 182.—A FORTIFIED TOWN BEING BOMBARDED BY A CATAPULT.

Criticism.—The stones thrown by the besieged may be seen falling in the trenches of the
besiegers.   The catapult depicted is drawn on much too small a scale.

From Polybius.   Edition of 1727.

Among the older authors quoted, Polybius, Ramelli, Valturius, Vegetius
and Vitruvius give illustrations of siege engines, those of Polybius, as supplied
by his commentator de Folard, being the most numerous.

Josephus gives an admirable account from personal knowledge of balistas and catapults in warfare, especially of their effects at the siege of Jotapata, A.D. 67, and at that of Jerusalem, A.D. 70. See pp. 267, 268.

Cæsar, Marcellinus, Plutarch and Tacitus also more or less fully describe these engines and the destruction they caused.

Among later writers, Père Daniel and Grose treat siege engines in considerable detail ; Grose giving many drawings of balistas and catapults.

Viollet-le-Duc in his exhaustive work on military architecture has several excellent illustrations of ancient siege engines, derived like those of Grose from the books and manuscripts of mediæval authors.

The late Emperor of the French, Napoleon III, was much interested in historic weapons. In an elaborate book he ordered to be compiled on military arms ancient and modern, entitled ' Etudes sur l'artillerie,' there are plans of the full-sized balista and trebuchet which he caused to be made, and with which many experiments were carried out in Paris to ascertain what were the effects of similar engines in ancient and mediæval warfare.

Some years ago these models were to be seen in the museum of Roman antiquities at Saint Germain-en-Laye, but I do not know if they are there now.

The largest siege engines used in ancient times were so ponderous that it was often impossible to transport them overland in bulk. For this reason, unless carriage by water was available, the principal parts of such an engine, as its winches, windlasses and cordage, were usually carted separately to the vicinity of the town about to be besieged. Its wooden framework was then made on the spot from trees cut down in the neighbourhood.

In some cases we read that the huge logs of wood which formed the frame of a heavy engine were dragged independently by oxen to the scene of attack and there fitted together.

FIG. 183.—A SIEGE.

Criticism.—The picture is open to the spectator in order that he may see both defenders and besiegers at work.
The besieged have just cast a stone from a catapult. The stone is falling on the movable tower belonging to the attacking side.
The catapult is, however, too small, and could not cast a stone of the size shown.

From Polybius. Edition of 1727.

CHAPTER LII

THE ANTIQUITY OF BALISTAS AND CATAPULTS

UNDER this heading I can but give quotations from various sources.

It is evident that a history of ancient siege engines cannot be created *de novo*. All that can be done is to quote with running criticism what has already been written about them. These remarks apply equally to the next chapter.

The first mention of balistas and catapults is to be found in the Old Testament, two allusions to these weapons being made therein.

The references are :

2 Chronicles xxvi. 15, 'And he[1] made in Jerusalem engines, invented by cunning men, to be on the towers and upon the bulwarks to shoot arrows and great stones withal.'

Ezekiel xxvi. 9, 'And he shall set engines of war against thy walls.'

Though the latter extract is not so positive in its wording as the one first given, it undoubtedly refers to engines that cast either stones or arrows against the walls, especially as the prophet previously alludes to other means of assault.

One of the earliest and most authentic descriptions of the use of great missive engines, is to be found in the account by Plutarch of the siege of Syracuse by the Romans, 214–212 B.C. See pp. 265, 266.

Cæsar in his Commentaries on the Gallic and Civil wars, B.C. 58–50, frequently mentions the engines which accompanied him in his expeditions.

The balistas on wheels were harnessed to mules and called carro-balistas.

The carro-balista discharged its heavy arrow over the head of the animal to which the shafts of the engine were attached. Among the ancients, these carro-balistas acted as field artillery and one is plainly shown in use on Trajan's Column.

According to Vegetius, every cohort was equipped with one catapult and every century with one carro-balista ; eleven soldiers being required to work the latter engine.

[1] Uzziah.

Sixty carro-balistas accompanied, therefore, a legion, besides ten catapults. The catapults were drawn along with the army on great carts yoked to oxen.

In the battles and sieges sculptured on Trajan's Column there are several figures of balistas and catapults. This splendid monument was erected in Rome, 105–113, to commemorate the victories of Trajan over the Dacians,

FIG. 184.—A SMALL CATAPULT ON WHEELS FOR USE AS LIGHT FIELD ARTILLERY.

Criticism.—The arm is here wound down and held by its catch, but there are no winches shown for twisting the skein of cord between which the arm works, and, in fact, no space for them.

From Polybius. Edition of 1727.

and constitutes a pictorial record in carved stone containing some 2,500 figures of men and horses.

In nearly every siege of note recorded in history, from that of Syracuse (214–212 B.C.) until the end of the fourteenth century—balistas or catapults are mentioned.

The trebuchet does not, however, appear to have been common in warfare before the middle of the twelfth century.

It is astonishing what a large number of catapults and balistas were sometimes used in a siege. For instance, at the conquest of Carthage, B.C. 146, 120 great catapults and 200 small ones were taken from the defenders, besides 33 great balistas and 52 small ones (Livy).[1]

Abulfaragio (Arab historian 1226–1286) records that at the siege of Acre in 1191, 300 catapults and balistas were employed by Richard I. and Philip II.

Abbo, a monk of Saint Germain des Prés, in his poetic but very detailed account of the siege of Paris by the Northmen in 885, 886, writes 'that the besieged had a hundred catapults on the walls of the town.'[2]

Among our earlier English kings, Edward I. was the best versed in projectile weapons large and small, including crossbows and longbows.

In the Calendar of Documents relating to Scotland, an account is given of his 'War-wolf,' a siege engine in the construction of which he was much interested.

This machine was of immense strength and size and took fifty carpenters and five foremen a long time to complete. Edward designed it for the siege of Stirling, whither its parts were sent by land and by sea.

Sir Walter de Bedewyne writing to a friend on July 20, 1304 (see Calendar of State Documents relating to Scotland), says: 'As for news, Stirling Castle was absolutely surrendered to the King without conditions this Monday, St. Margaret's day, but the King wills it that none of his people enter the castle till it is struck with his "War-wolf," and that those within the castle defend themselves from the said "War-wolf" as best they can.'

From this it is evident that Edward having constructed his 'War-wolf' to cast heavy stones into the castle of Stirling to induce its garrison to surrender, was much disappointed by their capitulation before he had an opportunity of testing the power of his new weapon.

Edward was not, however, to be baulked in this way, for he was anxious to try his 'War-wolf,' which had been transported to Stirling at much trouble and expense. For this reason he would not accept the surrender of the castle till he had shot off his 'War-wolf' at it, to see how the machine acted in warfare.

One of the last occasions on which the ancient form of siege engine was used with success, is described by Guillet in his Life of Mahomet II.[3] This

---

[1] Just previous to the famous defence of Carthage, the Carthaginians surrendered to the Romans 'two hundred thousand suits of armour and a countless number of arrows and javelins, besides catapults for shooting swift bolts and for throwing stones, to the number of two thousand.' From Appian of Alexandria, a Greek writer who flourished 98–161.

[2] These were probably balistas, as Ammianus Marcellinus writes of the catapult, 'An engine of this kind placed on a stone wall shatters whatever is beneath it, not by its weight but by the violence of its shock when discharged.'

[3] Guillet de Saint George, born about 1625, died 1705. His *Life of Mahomet II* was published in 1681. He was the author of several other works, including one on riding, warfare and navigation termed the *Gentleman's Dictionary*. The best edition of this book is in English and has many very curious illustrations. It is dated 1705.

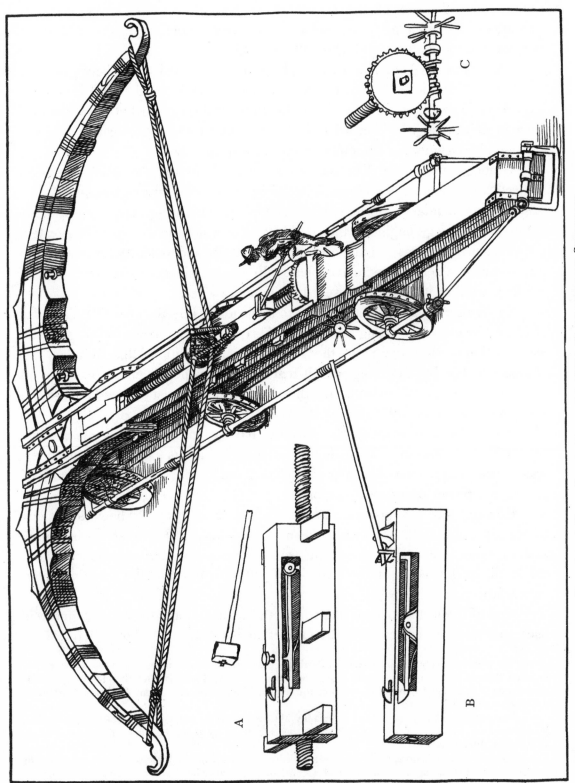

FIG. 185.—A SIEGE CROSSBOW IN THE FORM OF AN IMMENSE STONEBOW.

From 'Il Codice Atlantico,' Leonardo da Vinci, 1445-1520. For criticism, see page 263.

author writes : 'At the siege of Rhodes in 1480, the Turks set up a battery of sixteen great cannon, but the Christians successfully opposed the cannon with a counter-battery of new invention.[1]

'An engineer, aided by the most skilful carpenters in the besieged town, made an engine that cast pieces of stone of a terrible size. The execution wrought by this engine prevented the enemy from pushing forward the work of their approaches, destroyed their breastworks, discovered their mines and filled with carnage the troops that came within range of it.'

At the siege of Mexico by Cortes in 1521, when the ammunition for the Spanish cannon ran short, a soldier with a knowledge of engineering undertook to make a trebuchet that would cause the town to surrender. A huge engine was constructed, but on its first trial the rock with which it was charged instead of flying into the town, ascended straight upwards and falling back to its starting-point destroyed the mechanism of the machine itself.[2]

Though all the projectile engines worked by cords and weights disappeared from warfare when cannon came to the front in a more or less improved form, catapults—if Vincent le Blanc is to be credited—survived in barbaric nations long after they were discarded in Europe.

This author (in his travels in Abyssinia) writes 'that in 1576 the Negus besieged Tamar a strong town defended by high walls, and that the besieged had engines composed of great pieces of wood which were wound up by cords and screwed wheels and which unwound with a force that would shatter a vessel, this being the cause why the Negus did not assault the town after he had dug a trench round it.'[3]

---

[1] Called a new invention because the old siege engine of which this one (probably a trebuchet) was a reproduction had previously been laid aside for many years.

[2] *Conquest of Mexico.* W. Prescott, 1843.

[3] Vincent le Blanc, *Voyages aux quatre parties du monde, redigé par Bergeron*, Paris, 1649. Though the accounts given by this author of his travels are imaginative, I consider his allusion to the siege engine to be trustworthy, as he was not likely to invent so correct a description of a catapult.

FIG. 185.—*Criticism.*—A stonebow of vast size. A and B represent two kinds of lock. In A, the catch of the lock over which the loop of the bow-string was hitched, was released by striking down the knob to be seen below the mallet. In B, the catch was set free by means of a lever. C, shows the manner of pulling back the bow-string. By turning the spoked wheels, the screw-worm revolved the screwed bar on which the lock A, travelled. The lock, as may be seen, worked to or fro in a slot along the stock of the engine. In the illustration the bow is fully bent and the man indicated is about to discharge the engine. After this was done, the lock was wound back along the screw-bar and the bow-string was hitched over the catch of the lock preparatory to bending the bow again.

Besides being a famous painter, Leonardo was distinguished as an inventor of and exact writer on mechanics and hydraulics.

'No artist before his time ever had such comprehensive talents such profound skill or so discerning a judgment to explore the depths of every art or science to which he applied himself.'

JOHN GOULD, *Dictionary of Painters*, 1839.

From the above eulogy we may conclude that the drawings of ancient siege engines by Leonardo da Vinci are fairly correct.

The final appearance of the catapult in warfare occurred during the siege of Gibraltar by the French and Spanish fleets, 1779–82.

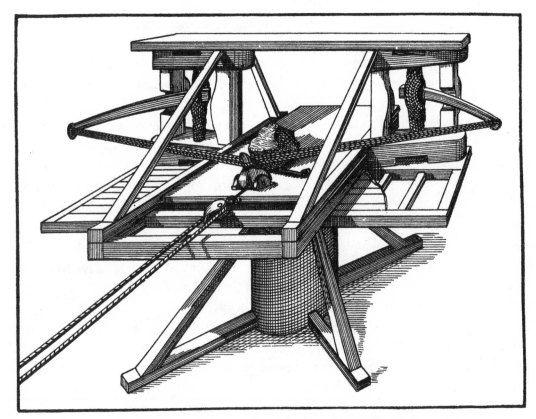

FIG. 186.—A BALISTA FOR THROWING LARGE STONES.

Criticism.—A balista of this kind, in my opinion, was never used for throwing stones.
This weapon certainly could not do so unless for a short distance along the ground.
Its bow-string would soon be cut to pieces and there is no means of projecting the missile upwards.

From Vegetius. Edition of 1607.

On this occasion General Melville,[1] at the desire of Lord Heathfield,[2] caused a catapult to be made for projecting heavy stones over the edge of the precipice, so that they might fall on a ledge of rock frequented by the Spaniards where shells from mortars could not be aimed to reach them.

[1] General Robert Melville, soldier and antiquary, born 1723, died 1809. He invented the carronade for use on board ships of war.
[2] General George Eliot, created Lord Heathfield, 1787, for his successful defence of Gibraltar against the French and Spanish fleets, 1779-1782. His portrait by Sir Joshua Reynolds is in the National Gallery, London.

CHAPTER LIII

THE EFFECTS OF ANCIENT SIEGE ENGINES IN WARFARE

PLUTARCH in his life of Marcellus the Roman General, gives a graphic account of Archimedes and the engines this famous mathematician employed in the defence of Syracuse.

It appears that Archimedes showed his relative Hiero II, King of Syracuse, some wonderful examples of the way in which immense weights could be moved by a combination of levers.

Hiero being greatly impressed by these experiments, entreated Archimedes temporarily to employ his genius in designing articles of practical use, with the result that the scientist constructed for the King all manner of engines suitable for siege warfare.

Though Hiero did not require the machines, his reign being a peaceful one, they proved of great value shortly after his death when Syracuse was besieged by the Romans under Marcellus, 214–212 B.C.

On this occasion Archimedes directed the working of the engines he had made some years previously for Hiero.

Plutarch writes : ' And in truth all the rest of the Syracusans were no more than the body in the batteries of Archimedes, whilst he was the informing soul. All other weapons lay idle and unemployed, his were the only offensive and defensive arms of the city.'

When the Romans appeared before Syracuse its citizens were filled with terror, for they imagined they could not possibly defend themselves against so numerous and fierce an enemy.

But, Plutarch tells us, ' Archimedes soon began to play his engines upon the Romans and their ships, and shot against them stones of such an enormous size and with so incredible a noise and velocity that nothing could stand before them. The stones overturned and crushed whatever came in their way and spread terrible disorder through the Roman ranks. As for the machine which Marcellus brought upon several galleys fastened together, called *sambuca* [1]

---

[1] *Sambuca.* A stringed instrument with cords of different lengths like a harp. The machine which Marcellus brought to Syracuse was designed to lift his soldiers—in small parties at a time and in quick succession—over the battlements of the town, so that when their numbers inside it were sufficient they might open its gates to the besiegers. The soldiers were intended to be hoisted on a platform, worked up

from its resemblance to the musical instrument of that name; whilst it was yet at a considerable distance, Archimedes discharged at it a stone of ten talents' weight and after that a second stone and then a third one, all of which striking it with an amazing noise and force completely shattered it.[1]

Marcellus in distress drew off his galleys as fast as possible and sent orders to his land forces to retire likewise. He then called a council of war, in which it was resolved to come close up to the walls of the city the next morning before daybreak, for they argued that the engines of Archimedes being very powerful and designed to act at a long distance, would discharge their projectiles high over their heads. But for this Archimedes had been prepared, for he had engines at his disposal which were constructed to shoot at all ranges. When therefore the Romans came close to the walls, undiscovered as they thought, they were assailed with showers of darts, besides huge pieces of rock which fell as it were perpendicularly upon their heads, for the engines played upon them from every quarter.

'This obliged the Romans to retire, and when they were some way from the town Archimedes used his larger machines upon them as they retreated, which made terrible havoc among them as well as greatly damaged their shipping. Marcellus, however, derided his engineers and said, " Why do we not leave off contending with this geometrical Briareus, who sitting at ease and acting as if in jest has shamefully baffled our assaults, and in striking us with such a multitude of bolts at once exceeds even the hundred-handed giant of fable?"

'At length the Romans were so terrified that if they saw but a rope or a beam projecting over the walls of Syracuse, they cried out that Archimedes was levelling some machine at them and turned their backs and fled.'

As Marcellus was unable to contend with the machines directed by Archimedes, and as his ships and army had suffered severely from the effects of these stone- and javelin-casting weapons, he changed his tactics and instead of besieging the town he blockaded it and finally took it by surprise.

Though at the time of the siege of Syracuse, Archimedes gained a reputation of divine rather than human knowledge in regard to the methods

and down by ropes and winches. As the machine was likened to a harp, it is probable it had a huge curved wooden arm fixed in an erect position and of the same shape as the modern crane used for loading vessels. If the arm of the *sambuca* had been straight like a mast, it could not have swung its load of men over a wall. Its further resemblance to a harp would be suggested by the ropes which were employed for lifting the platform to the summit of the arm, these doubtless being fixed from the top to the foot of the engine.

[1] It is I consider impossible that Archimedes, however marvellous the power of his engines, was able to project a stone of ten Roman talents or nearly 600 lbs. in weight, to a considerable distance ! Plutarch probably refers to the talent of Sicily, which weighed about 10 lbs. A stone of ten Sicilian talents, or say 100 lbs., could have been thrown by a catapult of great strength and size.

Though the trebuchet cast stones of from 200 lbs. to 300 lbs. and more this weapon was not invented till long after the time of Archimedes.

he employed in the defence of the city, he left no description of his wonderful engines, for he regarded them as mere mechanical appliances which were beneath his serious attention, his life being devoted to solving abstruse questions of mathematics and geometry.

Archimedes was slain at the capture of Syracuse, B.C. 212, to the great

FIG. 187.—A SIEGE CATAPULT.

Criticism.—This is an excellent representation of a catapult, though the engine is small for siege use. In this case the arm of the weapon is wound down till its point is secured by the catch to be seen beneath it (see fig. 201, p. 295, for this kind of lock).

One end of the winding rope is made fast to an iron bar fixed across the framework of the machine and its other end to the roller the men are turning.[1]   There is, however, no indication of how the arm was cast off from the rope when the former was wound down to the catch, nor is there any form of safety check to be seen on the winding roller.

From Polybius. Edition of 1727.

regret of Marcellus.   Of this event Plutarch gives several versions, one of which is that 'a soldier suddenly entered his room and ordered him to follow him to Marcellus, but Archimedes refusing to do this till he had completed the problem he was engaged on, the soldier drew his sword and killed him.'

The following extracts from Josephus, as translated by Whiston, enable us to form an excellent idea of the effects of great catapults in warfare :

(1) *Wars of the Jews*, Book III., Chapter VII.—The siege of Jotapata, A.D. 67. 'Vespasian then set the engines for throwing stones and darts round

---

[1] This reduced by one half the strain of winding down the arm.   See remarks, p. 296.

about the city ; the number of the engines was in all a hundred and sixty. . . . At the same time such engines as were intended for that purpose, threw their spears buzzing forth, and stones of the weight of a talent were thrown by the engines that were prepared for doing so. . . .

'But still Josephus and those with him, although they fell down dead one upon another by the darts and stones which the engines threw upon them, did not desert the wall. . . . The engines could not be seen at a great distance and so what was thrown by them was hard to be avoided ; for the force with which these engines threw stones and darts made them wound several at a time, and the violence of the stones that were cast by the engines was so great that they carried away the pinnacles of the wall and broke off the corners of the towers ; for no body of men could be so strong as not to be overthrown to the last rank by the largeness of the stones. . . . The noise of the instruments themselves was very terrible, the sound of the darts and stones that were thrown by them was so also ; of the same sort was that noise the dead bodies made when they were dashed against the wall.'

(2) *Wars of the Jews*, Book V., Chapter VI.—The siege of Jerusalem, A.D. 70. 'The engines that all the legions had ready prepared for them were admirably contrived ; but still more extraordinary ones belonged to the tenth legion : those that threw darts and those that threw stones were more forcible and larger than the rest, by which they not only repelled the excursions of the Jews but drove those away who were upon the walls also. Now the stones that were cast were of the weight of a talent [1] and were carried two or more stades.[2]

'The blow they gave was no way to be sustained, not only by those who stood first in the way but by those who were beyond them for a great space.

'As for the Jews, they at first watched the coming of the stone for it was of a white colour, and could therefore not only be perceived by the great noise it made but could be seen also before it came by its brightness ; accordingly the watchmen that sat upon the towers gave notice when an engine was let go . . . so those that were in its way stood off and threw themselves down upon the ground. But the Romans contrived how to prevent this by blacking the stone ; they could then aim with success when the stone was not discerned before-hand, as it had been previously.'

The accounts given by Josephus are direct and trustworthy evidence, for the reason that this chronicler relates what he personally witnessed during the sieges he describes, in one of which (Jotapata), he acted the part of a brave and resourceful commander.

Tacitus in describing a battle fought near Cremona between the armies

---

[1] 57¾ lbs. (avoirdupois).
[2] Two stades would be 404 yards ; the measure of a stade is 606¾ English feet.

of Vitellius and Vespasian, A.D. 69, writes : 'The Vitellians at this time changed the position of their battering-engines, which in the beginning were

FIG. 188.—A STATIONARY BALISTA FOR USE IN A SIEGE.

Criticism.—The skeins of cordage between which the arms of the engine work are apparently cut off at their centres, and no provision for holding the lower winches is shown. However great the power of this engine, it could merely discharge its javelin some 50 yards along the ground, as there is no means of elevating the groove in which the arrow is laid.

From Polybius. Edition of 1727.

placed in different parts of the field and could only play at random against the woods and hedges that sheltered the enemy. They were now moved to the Posthumian way, and thence having an open space before them could discharge their missiles with good effect.'[1]

[1] Tacitus continues : 'The fifteenth legion had an engine of enormous size, which was played off with dreadful execution and discharged massy stones of a weight to crush whole ranks at once. Inevitable ruin

Froissart chronicles that at the siege of Thyn-l'Evêque, 1340, in the Low Countries, ' John, Duke of Normandy had a great abundance of engines carted from Cambrai and Douai. Among others he had six very large ones which he placed before the fortress, and which day and night cast great stones which battered in the tops and roofs of the towers and of the rooms and halls, so much so that the men who defended the place took refuge in cellars and vaults.'

Camden records that the strength of the engines employed for throwing stones was incredibly great and that with the engines called mangonels[1] they used to throw millstones. Camden adds that ' when King John laid siege to Bedford Castle there were on the east side of the castle two catapults battering the old tower, as also two upon the south side besides another on the north side which beat two breaches in the walls.'

The same authority asserts that when Henry III. was besieging Kenilworth Castle the garrison had engines which cast stones of an extraordinary size, and that near the castle several balls of stone sixteen inches in diameter have been found, which are supposed to have been thrown by engines with slings[2] in the time of the Barons' war.

Holinshed writes that ' when Edward I. attacked Stirling Castle, he caused an engine of wood to be set up to batter the castle which shot stones of two or three hundredweight,' p. 261.

Père Daniel in his *Histoire de la Milice Françoise*, writes : ' The great object of the French engineers was to make siege engines of sufficient strength to project stones large enough to crush in the roofs of houses and break down the walls.' This author continues : ' The French engineers were so successful and cast stones of such enormous size that their missiles even penetrated the vaults and floors of the most solidly built houses.'[3]

The effects of the balista on the defenders of a town were in no degree inferior to those of the catapult. The missile of the balista consisted of a huge steel-tipped wooden bolt which, although of far less weight than the great ball of stone cast by a catapult or the far larger one thrown by a trebuchet, was able to penetrate roofs and destroy light parapets. Cæsar records that ' when his lieutenant Caius Trebonius was building a movable tower at the siege of Marseilles, the only method of protecting the workmen from the darts of engines[4]

---

must have followed if two soldiers had not signalised themselves by a brave exploit. Covering themselves with shields of the enemy which they found among the slain, they advanced undiscovered to the battering-engine and cut its ropes and springs. In this bold adventure they both perished and with them two names that deserved to be immortal.'

[1] Catapults were often called mangons or mangonels in early mediæval warfare, but in course of time the name mangonel was applied to any siege engine that projected stones or arrows. In this case the trebuchet is intended as no catapult could project a millstone.

[2] The engines here alluded to by Camden were trebuchets as the catapult had not a sling attached to its arm.   [3] These engines would also be trebuchets.   [4] Balistas.

was by hanging curtains woven from cable-ropes on the three sides of the tower exposed to the besiegers.[1]

Procopius relates that during the siege of Rome in 537 by Vitiges King of Italy, he saw a Gothic chieftain in armour suspended to a tree which he had climbed, and to which he had been nailed by a balista bolt which had passed through his body and then penetrated into the tree behind him.

Again, at the siege of Paris by the Northmen in 885–886, Abbo writes

FIG. 189.—CASTING A DEAD HORSE INTO A BESIEGED TOWN BY MEANS OF A TREBUCHET.
From ' Il Codice Atlantico,' Leonardo da Vinci, 1445-1520.

that Ebolus[2] discharged from a balista a bolt which transfixed several of the enemy.

With grim humour Ebolus bade their comrades carry the slain to the kitchen, his suggestion being that the men impaled on the shaft of the balista resembled fowls run through with a spit previous to being roasted.

Not only were ponderous balls of stone and heavy bolts projected into a town and against its walls and their defenders, but with a view to causing

---

[1] 'For this was the only sort of defence which they had learned by experience in other places could not be pierced by darts or engines.' Cæsar's *Commentaries on the Civil War*, Book ii., Chapter ix.

[2] Abbot of Saint-Germain des Prés and one of the chief defenders of the town.

a pestilence it was also the custom to throw in dead horses and even the bodies of soldiers who had been killed in sorties or assaults.

For example, Varillas[1] writes that 'at his ineffectual siege of Carolstein in 1422, Coribut caused the bodies of his soldiers whom the besieged had killed to be thrown into the town, in addition to 2,000 cartloads of manure. A great number of the defenders fell victims to the fever which resulted from the stench, and the remainder were only saved from death by the skill of a rich apothecary who circulated in Carolstein remedies against the poison which infected the town.'[2]

Froissart tells us that at the siege of Auberoche, an emissary who came to treat for terms was seized and shot back into the town. This author writes:

'To make it more serious, they took the varlet and hung the letters round his neck and instantly placed him in the sling of an engine and then shot him back again into Auberoche. The varlet arrived dead before the knights who were there and who were much astonished and discomfited when they saw him arrive.'

Another historian explains that to shoot a man from the sling of an engine he must first be tied up with ropes, so as to form a round bundle like a sack of grain.

The engine with which such fiendish deeds were achieved was the trebuchet.

A catapult was not powerful enough to project the body of a man. This difficulty was overcome by cutting off the head of any unfortunate emissary for peace, if the terms he brought were scornfully rejected. His letter of supplication from the besieged was then nailed to his skull, and his head was sent flying through space to fall inside the town as a ghastly form of messenger conveying a refusal to parley.

As it was always an object to the besiegers of a town to start a conflagration if they could, Greek fire was used for the purpose. The flame of this fearfully destructive liquid, the composition of which is doubtful, could not be quenched by water. It was placed in round earthenware vessels that broke on falling and which were shot from catapults; as the roofs of mediæval dwelling-houses were usually thatched, it of course dealt destruction when it encountered such combustible material.

To conclude this chapter I give some quaint extracts from de Folard, who supplied the commentary to Vincent Thuillier's translation of Polybius printed in 1727. de Folard was a soldier and a writer on military tactics.

---

[1] French historian, born 1624, died 1696.

[2] The rebels of Bohemia, the Hussites, first offered the crown of Bohemia to Jagellon King of Poland, who declined it. They then offered it to the Grand Duke of Lithuania, who accepted it. Prince Coribut besieged Carolstein as a General acting on behalf of the Grand Duke of Lithuania.

His recommendation of the Roman siege engine in preference to the cannon of his time reads very curiously, especially when published at so late a date as 1727.

His reference to the crossbow as a superior arm to the gun in warfare is also interesting, however primitive the latter weapon may have then been.

de Folard writes : ' I am convinced that if it were possible to lay aside the prejudice of custom, catapults and balistas would soon reduce to silence our mortars and swivel-guns ; for who can doubt that catapults were the more useful for throwing stones and bombs ? How much would they not save in sieges ?

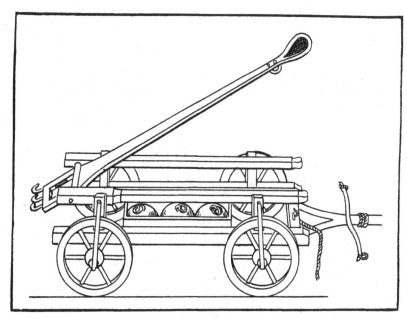

FIG. 190.—A TREBUCHET ON WHEELS. THE DETACHABLE IRON WEIGHTS WHICH HOOK ON TO THE BUTT OF THE ARM MAY BE SEEN INSIDE THE FRAMEWORK OF THE ENGINE.

Criticism.—The arm has no sling attached to it and resembles the arm of a catapult.

An engine of this construction would not be able to cast a stone 100 yards, if, indeed, 50 yards.

From Père Daniel. Edition of 1721

What paraphernalia, what an array of material, equipages, horses, men and workmen for the service and transport of mortars !

' It is apparent that a dozen mules sufficed for the carriage of the ropes, winches, arms, cushions and all the utensils necessary for several large catapults. All else required for their construction can be found wherever there are trees of a certain size, no matter what the nature of the wood.

' All kinds of stones are suitable for catapults, whereas balls only can be used for our swivel-guns. A mortar of the greatest capacity can scarcely throw a weight of 60 lbs., while the catapult can throw 100 lbs.

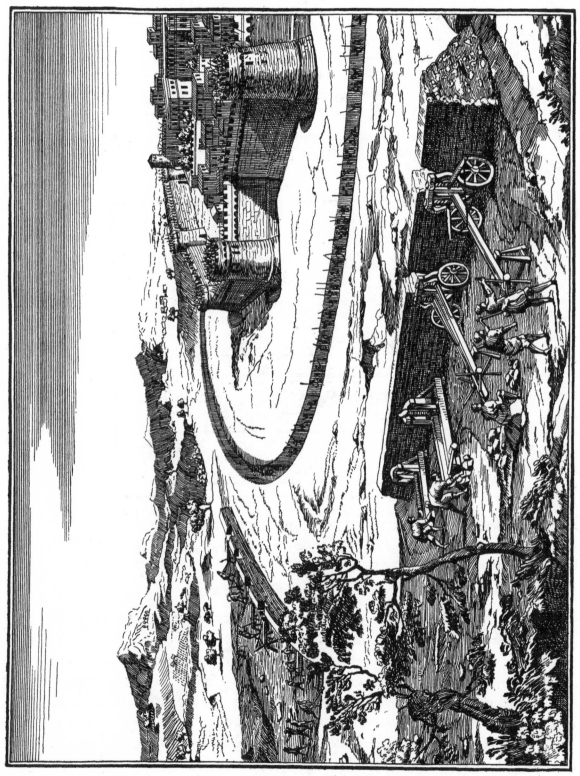

FIG. 191.—BESIEGING A FORTIFIED TOWN WITH A BATTERY OF CATAPULTS AND BALISTAS.

Criticism.—In this picture the balistas are fairly correct, but the catapults are too small.

From Polybius. Edition of 1727.

'What is most to be valued in a catapult is the certainty of its effect and the directness of its aim. One can rely upon shooting the stones to the desired point, for there is no cause which can make a catapult shoot to a greater or a less distance, or to aim at one time more or less exactly than at another when it is at the same degree of elevation and tension. It is not so with our mortars, on account of the different effects or qualities of the powder : for although powder is the same in appearance it is not so in its effects. One barrel is never the same as another barrel. Powder is never just the same in quality and strength.

'The catapult has, besides, an infinite number of advantages over mortars. We can by different inclinations place a catapult so that its stone can be thrown exactly where we wish, which is what we cannot be sure of doing with our mortars.

'Another very great advantage is that catapults and such engines make no noise. Now, one is warned by the noise of the mortars and can pretty well judge what the swivel-gun is about; but by night or by day one cannot tell whether the catapult has or has not shot off its stone.

'This kind of engine is even more advantageous in a besieged place. It is not necessary that it should be placed on the ramparts, and it is rarely that it need be in a position to receive the shots aimed at it. Besides this, the besieged are able to have larger engines than the besiegers and to throw weights of enormous bulk that are able to break the galleries and greatly retard the building of trenches, which drives the besiegers to infinite precaution and compels them to continue their work from under cover.

'The crossbow was infinitely deadly and more effective than our gun; its aim more certain and precise and its strength at least equal. If we had not introduced the bayonet at the end of the gun, which constitutes almost all the advantage of this arm, the crossbow would have far surpassed it. I do not, however, wish to treat the gun with contempt, as it is of utility in combats where it is impossible to have a large area as in assaults and offensive attacks.'

CHAPTER LIV

THE DISTANCES TO WHICH ANCIENT SIEGE ENGINES CAST THEIR PROJECTILES

THE catapults, balistas and trebuchets employed for bombarding the walls, houses and people of a town, were, of course, placed well out of range of the bows and crossbows of its defenders.

If the besiegers located their engines within reach of arrows, the men who worked the engines would be slain by the archers of the opposing side, especially as it was not possible to shelter the larger machines, such as the trebuchets, behind screens of wood or earth on account of their great size and height.

With the advantage of shooting downwards from the commanding elevation of towers and battlements, the archers were certainly able to attain a range of from 270 to 280 yards, and in any case could shoot considerably farther than they were able to do when standing on level ground.

In order merely to ensure their safety from archers, it would, therefore, be necessary to place the engines at about 300 yards from the outer walls of a besieged town.

As catapults were not only required to hurl their missiles against the towers and battlements of a town, but were designed also to shoot clear over the walls upon the houses and soldiers inside the defences, it is evident that whether large or small they must have had a range of from 350 to 400 yards to be effective.[1] See extracts from Josephus, p. 268.

Which side could produce the larger and more powerful engines was always an important point among the combatants at a siege, the advantage at first being usually with the besieged, as they could build their engines in time of peace and keep them ready for war. On the other hand, the besiegers had to bring their smaller engines from a distance and, as was usual, construct their larger ones on the spot.

---

[1] 400 yards was an immense distance for even a 50-lb. stone to be projected by a weapon that derived its power merely from twisted cordage. 450 yards was probably the extreme range of any of these engines.

The successful attack or defence of a fortified town often depended there-fore on which of the armies engaged had the more powerful balistas, catapults or trebuchets, as one engine of superior range could work destruction un-impeded if it happened that a rival of similar power was not available to check its depredations.

Froissart relates that 'at the siege of Mortagne in 1340, an engineer within the town constructed an engine to keep down the discharges of one

FIG. 192.—A SIEGE CATAPULT.

Criticism.—An excellent drawing of a catapult. This engine was moved into position on rollers and then props were placed under its sides to adjust the range of the projectile.

The end of the arm was secured by the notch of the large iron catch and was released by striking down the handle of the catch with a heavy mallet.

The arm is, however, too long for the height of the crossbar against which it strikes and would probably break off at its centre.

The hollow for the stone is much too large as a stone big enough to fit it could not be cast by a weapon of the dimensions shown in the picture.

From an Illustrated Manuscript, Fifteenth Century (No. 7239), Bibl. Nat. Paris.

powerful machine in the besieging lines. At the third shot he was so lucky as to break the arm of the attacking engine.' The account of this incident, as given by Froissart, is so quaint and graphic that I quote it here: 'The same day they of Valencens raysed on their syde a great engyn and dyd cast in stones so that it troubled sore them within the town. Thus yᵉ firstᵉ day passed and the night in assayling, and devysing how they might greve them in the fortress.

'Within Mortagne there was a connying maister in making of engyns who saw well how the engyn of Valencens did greatly greve them: he raysed an engyn in yᵉ castle, the which was not very great but he trymmed it to a point, and he cast therwith but three tymes. The firste stone fell a xii[1] fro the engyn without, the second fell on yᵉ engyn, and the thirde stone hit so true that it brake clene asonder the shaft of the engyn without; then the soldyers of Mortagne made a great shout, so that the Hainaulters could get nothing ther[2]; then the erle[3] sayd how he wolde withdrawe.'

(From the translation made at the request of Henry VIII. by John Bourchier second Lord Berners, published 1523–1525.)

These siege engines when only of moderate size were not always successful, as in some cases the walls of a town were so massively built that the projectiles of the enemy made little impression upon them. Froissart tells us that it was then the habit of the defenders of the walls to pull off their caps or produce cloths, and derisively dust the masonry when it was struck by stones.

With regard to the range of catapults, balistas and trebuchets many extravagant statements have been made by historians. François de Mézeray even declares that a catapult could shoot to a distance of a thousand yards![4]

On this point I have carefully sifted the evidence to be found in ancient and mediæval descriptions of sieges and have discarded all statements that are in the least doubtful.[5]   The conclusions I have arrived at will be found in the three following chapters and may, I am confident, be relied on as accurate.

---

[1] A foot.   [2] Could not throw any more stones.
[3] Count of Hainault.   He was besieging Tournay, but left that place and went to besiege Mortagne and ordered the people of Valenciennes to go with him.
[4] French historian; wrote a history of France in 3 vols., printed 1643–51.
[5] I have also had the advantage of possessing small and large working models from which to work out deductions and comparisons.

CHAPTER LV

THE CATAPULT, ITS CONSTRUCTION AND MANAGEMENT [1]

SURFACE VIEW OF THE FRAMEWORK OF A CATAPULT, FIG. 193, NEXT PAGE

I, II. The side-pieces. These are each 10 ft. 6 in. long and 1 ft. thick. They are 21 in. high at their forward ends in front of the skein and are reduced to a height of 15 in. at their after ends behind the skein; see fig. 194, p. 282, for a side view of the catapult.

III. The after cross-piece. This is 15 in. high and 1 ft. thick.
IV. The forward cross-piece. This is 21 in. high and 1 ft. thick.
The cross-pieces (III, IV) are cut into tenons at their extremities and mortised into the side-pieces I, II.

V. The small cross-piece (6 in. square). This gives additional support to the sides of the catapult to enable it to resist the immense force of the skein of twisted cord.

The inside width between the sides of the catapult (I, II), when the cross-pieces (III, IV, V) are fixed, is 4 ft.

A, A. The skein of twisted cord. [2] The ends of the skein turn over the crossbars of the large wheels (B, B), which twist the skein. See figs. 197, 199, pp. 286, 293.

C, C. The pinion wheels which turn the large wheels, B, B.

By turning with long spanners the spindle heads (D, D), of the pinion

[1] In connection with this chapter and the next one, refer to Frontispiece for sketches of a catapult being used.
[2] The holes in the sides of the catapult—through which the skein of cord passes to the winches—are, of course, of the same diameter as the apertures inside the large wheels of the winches. See fig. 193, and II, fig. 197, p. 286.

wheels (c, c), the large wheels (b, b), revolve and twist the skein of cord (a, a), between the halves of which the arm (e, e), is placed.[1]   See also Frontispiece.

The skein of cord (a, a), is 8 in. in diameter.

f, f. The roller (7 in. in diameter), which winds down the arm, e, e.

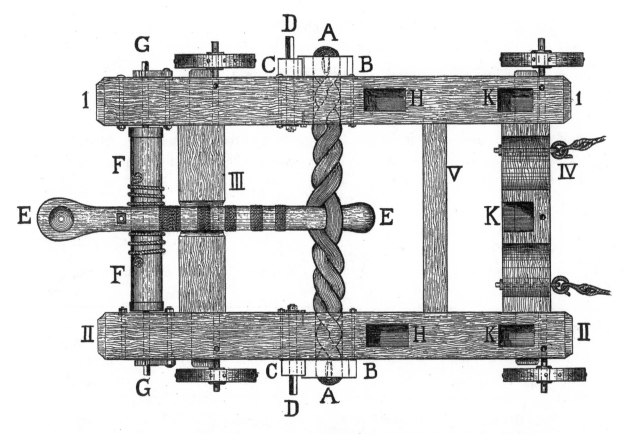

FIG. 193.—THE SURFACE VIEW OF THE FRAMEWORK, THE ARM[2] AND THE SKEIN OF TWISTED CORD OF A CATAPULT.   Scale ½ in. = 1 ft.

(To avoid confusion of details, the two uprights and the cross-bar between their tops against which the arm of the catapult strikes when released, are omitted in this plan.   They are shown in the other figures.)

The roller is revolved by two men, one on each side of the catapult.[3]   These men fit long spanners on the squared ends of the spindle, G, G.   This spindle passes through the centre of the roller, through the sides of the catapult and also through the four iron plates in which it revolves and which hold it in position. The plates are bolted to the sides of the catapult, fig. 193, and fig. 202, p. 298.

[1] The cogs in the wheels of the winches are omitted lest they should confuse other details.   These wheels are fully shown in fig. 197, p. 286.

[2] The arm is here wound down by the roller.   [3] See Frontispiece.

The two small cogged wheels—with their checks—which are fitted on the ends of the spindle (G, G), prevent the roller from reversing whilst the arm is being wound down, fig. 194, next page.

H, H. The mortises cut in the sides of the catapult to receive the tenons of the two uprights. Between the tops of these uprights is fixed the cross-bar against which the arm of the catapult rests, or when released from its catch strikes. The uprights and the cross-bar are shown in figs. 194, 195, 196, pp. 282, 284, 285.

It will be noticed that the mortises for the tenons of the uprights, are placed well away from the circular openings in the sides of the catapult through which the skein of cord passes. If these mortises were cut too near the openings for the skein, the side-pieces of the catapult would be weakened.

K, K, K. The mortises for the lower tenons of the three sloping supports which prevent the two uprights,—and their cross-bar,—from giving way under the blow of the released arm of the catapult, figs. 195, 202, pp. 284, 298.

The upper ends of the two side supports are mortised into the tops of the uprights, to which they are also bolted, fig. 194, next page, and fig. 202, p. 298.

The top of the middle support is mortised into the centre of the cross-bar that connects the uprights, fig. 195, p. 284, and fig. 202, p. 298.

SIDE VIEW OF THE CATAPULT, FIG. 194, NEXT PAGE

The arm (A) is here, ready to be wound down by the rope—$1\frac{1}{4}$ in. in diameter—that is attached to it and also to the roller. The ends of the rope are passed through holes in the winding roller and are then secured by knots, F, F, fig. 193.

The upper part or bend of the rope is hitched by a slip-hook to a ring-bolt which passes through the arm of the catapult. Fig. 200, p. 294, describes the ring-bolt and the slip-hook.

B. The position of the arm when it is fully wound down by the roller. The stone may be seen in the cup of the arm.

By pulling the cord (E), the arm is released from the slip-hook and—taking an upward sweep of 90 degrees (see curved line of arrows)—returns to its original position, as at A.

C. The position of the arm of the catapult at the moment when the stone leaves it. The stone is projected upwards at an angle of about 45 degrees, as represented by the straight line of small arrows that indicates its flight after it leaves the arm at C.

When the arm reaches the point in its upward sweep at which its speed is greatest, the stone instantly flies away in front of it.

That is to say, when the arm decreases in speed, however slightly, it cannot keep pace with the stone it projected the moment it reached its maximum velocity.

This principle should apply equally to the bow and its arrow. In this case I believe the arrow leaves the bow-string before the latter has returned to

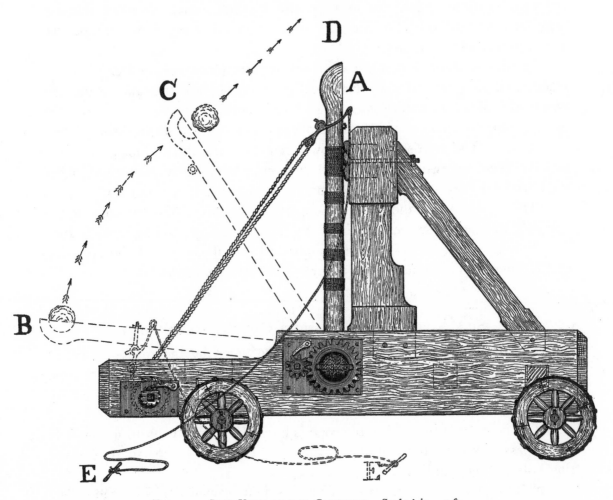

FIG. 194.—SIDE VIEW OF THE CATAPULT. Scale ½ in. = 1 ft.

its position of rest, or as it was before it was pulled back by the archer to discharge the arrow.[1]

When I originally directed my attention to the construction of a catapult,

[1] This theory regarding the bow and its arrow is of course suggested by the very evident action of the catapult. It is a theory, however, that is difficult to reduce to fact by eyesight or experiment. In the case of the catapult, the stone may be seen flying through the air before the sound of the arm striking the cross-piece is heard.

I concluded that the mediæval drawings which depicted the arm of the engine in a perpendicular position, as in A, fig. 194, were incorrect.

My surmise was that a catapult with a perpendicular arm would merely bowl its stone along the ground, on the principle that the stone was retained in the cup of the arm till the latter was checked by the cross-bar.

Carrying out this idea, I placed the winches of the first catapult I made in front of the uprights and not behind them as in the weapon here described.

By this arrangement the arm when released had of course an upward inclination when checked by its cross-bar. Such a position for example as half-way between C and A, fig. 194.

The result of this intended improvement on the ancients was,

WITH A SLOPED ARM

1. The cross-bar which checked the arm of the catapult was soon knocked loose through being struck in an upward direction.

2. The range of the projectile was unsatisfactory through the arm being wound down only a short distance from its state of rest.

3. The projectile—as in the case of a perpendicular arm—left its cup a considerable time before the arm encountered the cross-bar.

On the other hand I found that:

WITH A PERPENDICULAR ARM (A, FIG. 194)

1. The cross-bar was struck a level blow, or one that was taken by the three supports which lean against its centre and ends.

2. The range of the projectile was much increased owing to the additional distance the arm was wound down, and which caused the skein of cord to be far more tightly twisted than it was when the arm rested against the cross-bar in a sloping position before it was pulled back.

3. The projectile left the cup of the arm as shown at C, fig. 194; and as it did with a sloped arm.

Fig. 195 shows the large front cross-piece (IV, fig. 193, p. 280), between the sides of the catapult, as well as the three supports that hold the uprights and the cross-bar from movement when the latter is violently struck by the released arm.[1]

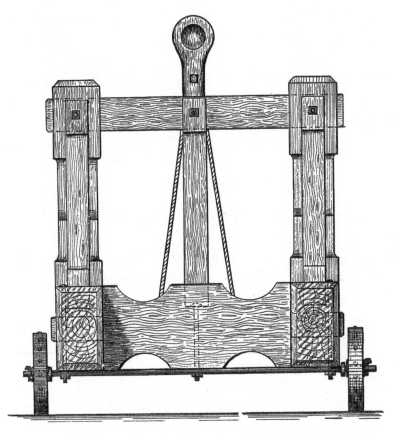

FIG. 195.—THE FRONT END OF THE CATAPULT.   Scale ½ in. = 1 ft.
The winches are here omitted.

Fig. 196, opposite page, shows the arm—the rope which pulls down the arm —the slip-hook for releasing the arm when it is wound down—the winding roller—the upper edge of the skein of cord—the winches—and the other parts of the engine previously described.

We also see in fig. 196 the padded cushion against which the arm strikes with terrific force when its upper end is checked by the cross-bar.   The cushion is of the same depth as the cross bar.   It is 16 in. long and about 6 in. thick.

[1] The top cross-bar, against which the arm strikes, should be of ash, 6 ft. 6 in. long, 8 in. square.   It should be reduced at its ends to 6 in. square and stepped into and bolted to the tops of the uprights

It is made of soft hide, doubled and packed with horsehair, and should be nailed to the cross-bar.

Without this protection the arm and cross-bar would soon be shattered.[1]

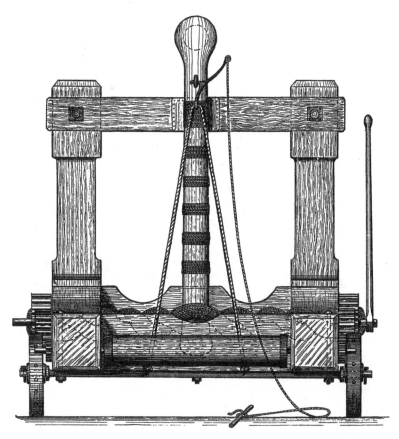

FIG. 196.—THE AFTER END OF THE CATAPULT. Scale ½ in. = 1 ft.

A spanner for turning the winches is shown in position on one of the pinion wheels.

THE ARM OF THE CATAPULT

The arm (of ash, straight grained and without a knot or shake) is 7 ft. long and 4½ in. thick, with rounded edges. It tapers from a width of 8 in. at its butt-end, to a width of 6½ in. at the part above the ring-bolt where it commences to enlarge into the cup that holds the stone.

The tendency of the arm of a catapult is always to draw out of the skein of cord, in which its butt-end is placed.

This is the result of the strain applied to the arm when it is being

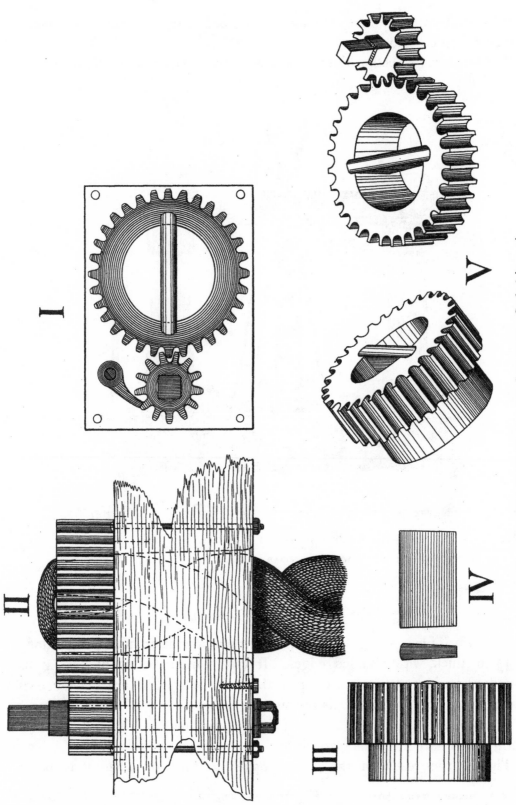

FIG. 197.—THE WINCHES OF THE CATAPULT.   Scale ⅛ in. = 1 in.

I. Surface view of one of the winches and its plate.   II. Side view of a winch as fitted in the catapult, with one end of the skein in position over the cross-bar of the large wheel of the winch.   III. Side view of the large wheel of a winch.   IV. The cross-bar of a winch.   V. Perspective views of the large wheel and pinion wheel of a winch.

wound down by the roller. To prevent this slipping of the arm its butt-end should be slightly increased in bulk, as shown in fig. 193, p. 280.

The cup or circular hollow at the end of the arm—in which the stone is laid—is 5 in. wide and 2 in. deep at its centre.

The arm should be tightly bound at short intervals with lashings of quarter-inch cord, fig. 196, page 285. Sometimes an arm will endure the great strain applied to it from the first and show no sign of fracture, though it may bend not a little when it is wound down to its full extent.

It is, however, probable that the first arm or two tried in the catapult will give way, especially if too much initial pressure is put upon them.

The arm should be tested by degrees and only pulled down its full distance after several trials at shorter ones.

The ancients had the same difficulty in obtaining arms for their large catapults that I have experienced with smaller ones.[1]

For this reason their engineers constructed the arm of a catapult of three longitudinal pieces.

They first fastened three smooth and closely fitting planks together with glue and with small rivets ; then they shaped the planks, thus held together, into an arm of correct size and outline.

The arm, except its enlarged head-end, was next wrapped tightly round its entire length with several layers, one above the other, of strong linen soaked in glue, the linen being cut in strips about 3 in. wide.

Finally strong cord, also soaked in glue, was closely lashed over the linen from the butt-end of the arm to the cup for the stone.

The arm was made on the same principle as a carriage spring, or a longbow of several pieces, and was infinitely stronger and more elastic than one formed of solid wood.

THE WINCHES OF THE CATAPULT, FIG. 197. FOR DIMENSIONS SEE NEXT PAGE

These are the most important parts of the catapult, and generate its projectile force.

However carefully a catapult may be constructed, its effectiveness chiefly depends upon the two winches that twist the skein of cord in which its arm works.

The plans in fig. 197 show a winch and its cross-bar in various positions.

---

[1] I smashed six arms in succession in the first fairly large catapult I made before I could obtain one that lasted. The piece of selected ash used by a cart-builder for cutting a pair of shafts from can usually be made into an arm for a catapult.

In the catapult I am describing, the dimensions of each winch are :

LARGE WHEEL.—14 in. diameter across its top surface.

Its bore (*i.e.* the aperture for the skein of cord), 8 in. diameter.

Total length of the wheel, 8 in.

Length of its flange that fits through the iron plate, 3 in.

Thickness of the flange, $\frac{3}{4}$ in.

PINION WHEEL.—6 in. diameter.   Its length, 4 in.

The projecting ends of the spindles of the pinion wheels are each 2 in. square and 5 in. long.   On these ends heavy spanners are fitted for twisting up the skein of cord.   See Frontispiece.

The cross-bars fixed across the apertures of the large wheels, and over which the ends of the skein of cord pass, are each 10 in. in length, 4 in. deep and $1\frac{1}{4}$ in. wide across their tops.

They decrease to 1 in. in width at their lower edges and are, therefore, slightly sloped at their sides, as shown in IV, fig. 197, page 286.   These cross-bars fit like wedges, into the slots cut to receive them inside the large wheels of the winches, fig. 197.   They are rounded on their exposed edges so as not to fray the cord they hold and, of course, they equally divide the apertures of the wheels.

Though this was the method of fixing the cross-bars adopted by the ancients, I have had my winches cast with their cross-bars solid with their wheels and not as separate pieces.

The wrought-iron plates through which the flanges of the large wheels of the winches pass and on which the projecting rims of these wheels revolve, are each 1 in. thick.   These plates are bolted to the sides of the catapult, fig. 202, page 298.

The round shanks of the spindles of the pinion wheels (secured at their ends by washers and nuts), also pass through these plates as well as through the sides of the framework of the catapult, II, fig. 197.

An almost inconceivable strain can be applied to the skein of cord by four or five men turning the winches of the catapult, a strain so immense that no arm of serviceable dimensions could be made to withstand the force that would have to be applied to wind it down.

Some mediæval writers describe the devices formerly employed for reducing the friction created between the rims of the large wheels of the winches and the iron plates on which they revolve.

In the catapults I have made, I have not however found anything of the kind—such as ball bearings—necessary, other than plenty of grease inserted between frictional surfaces.

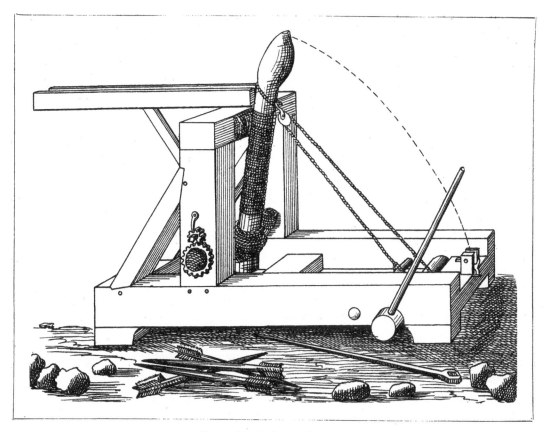

FIG. 198.—A SIEGE CATAPULT.

Criticism.—A catapult for discharging stones and javelins but an impossible engine in most respects.

In this case there is a grooved piece of wood (in the form of a shallow trough) on the top of the engine. This trough is intended to hold the javelin, the projecting butt-end of which is supposed to be struck by the released arm of the catapult. I do not believe a catapult was ever employed to project a javelin. It certainly could not do so in the manner here depicted, for the reason that the arm of the machine could never strike a true blow. Besides this, the arm of a catapult casts a stone with a slinging motion and does not recoil with the quick snap of a spring, such as would be necessary to flip a javelin forward and as is the case with the engine shown in fig. 216, p. 316.

Again, the winches for winding the skein of cordage are put in the weakest part of its framework in this catapult, *i.e.* between the uprights instead of between the sides where they should be.

From Polybius. Edition of 1727.

THE SKEIN OF CORD

We will now conclude that our catapult is ready for its skein of cord, its winches being in position one on each side of the framework.

In the first catapult I made I fitted a skein of thick rope for the arm to work between, but I found it was impossible to put an even strain upon the rope when twisting it up with the winches.

The result of this uneven strain was, that the lengths of rope which formed the skein—each $1\frac{1}{2}$ in. thick—broke one by one like rotten thread, owing to the force applied by the winches affecting them in detail instead of collectively.

After a series of experiments with various kinds of cordage, I discovered that the finer the cord used within reason, the more elastic and compact was the skein and hence the less its liability to break.

The fracture of a few strands of a large skein of fine cord is of no consequence, but the breaking of one stout rope amid a skein of a dozen lengths of such rope, means a noticeable loss of power.

The ancients were well aware of this and made the skeins of their catapults of thin cords of twisted hair.[1]

If horse-hair were not available in sufficient quantity, sinews from the necks of horses or oxen were used;[2] I do not find that ordinary rope was ever employed.

The elasticity of hair is so great, that however tight a large skein of it is twisted its extreme stretching or breaking limit cannot well be reached.

For this reason, there is always sufficient life or spring in the most tightly twisted skein of horse-hair to give the requisite velocity to the arm of the catapult.

It is evident that if the skein of a catapult were twisted up to its extreme limit, it would break under the further strain entailed on it by winding down the arm of the engine.

After testing every kind of material for the skein of a catapult I find that horse-hair rope—$\frac{1}{2}$ in. thick—is far the best.

Failing horse-hair, pure flax in the form of sailmaker's sewing twine is a fairly good substitute.

If this twine is used for the skein of a catapult it should be spun into a cord $\frac{1}{4}$ in. thick.

---

[1] In cases of emergency, woman's hair was made into skeins for catapults and balistas, and of all material nothing was so elastic or enduring for this purpose. When the inhabitants of Carthage commenced the heroic defence of their city (149–146 B.C.) they were forced to hurriedly manufacture weapons of all kinds to replace those which they had recently surrendered to the Roman general Censorinus (see footnote, p. 261). In various modern works we read of how 'the noble matrons of Carthage cut off their long tresses and twisted them into ropes for catapults.'

I can find no authority for any such picturesque writing, as ancient authors simply record the fact 'that women's hair was used at Carthage.' For instance, Florus, in his Roman History, a chronicler who flourished early in the second century, writes 'and the women parted with their hair to make cordage for the catapults.' Again, Zonaras, Byzantine historian, *Chronica*, ix. 26, says 'for the ropes of the catapults they used the hair of the women.'

At the siege of Salona by Marcus Octavius, one of Pompey's generals, the Roman women cut off their hair that it might be made into ropes for the engines of the besieged.—*Cæsar's Commentaries on the Civil War*, Book iii. Chapter ix.

[2] *Ligamentum colli*, also known as *Ligamentum nuchæ*, see note, p. 64.

CHAPTER LVI

THE CATAPULT, ITS CONSTRUCTION AND MANAGEMENT (Concluded)

HOW TO MAKE AND FIT THE SKEIN OF THE CATAPULT
FIG. 199, P. 293

INSERT a thin stick into the ground halfway between the winches. Place it upright inside the framework of the catapult. This stick will serve to keep the halves of the skein separate as it is being made, so that when it is completed the arm of the catapult may be placed in position without difficulty. Turn the winches till both their cross-bars are perpendicular to the ground and in line with the stick.

Next secure one end of the cord you are using for the skein to the corner of the cross-bar of one of the winches.

Pass the other end of the cord through the holes in the sides of the catapult and round the cross-bar of the opposite winch, and then back again over the bar of the first winch. Do this in regular rotation to and fro, first on one side of the stick then on the other. Be careful not to cross the lengths of cord as you pass them between the winches, but keep them individually straight, tight and regular and alternately on either side of the stick, A, fig. 199, p. 293.

Do not wrap the cord at haphazard round the cross-bars of the winches, but lay the turns regularly from one end of each cross-bar to its other end and then back again.

When a complete layer of cord is wrapped over a cross-bar, place on it a strip of paper 1 in. wide. By concealing the last layer the paper will show you how to proceed with the next.[1]

---

[1] The last few turns of the cord will have to be passed through the winches by the aid of a piece of stout wire with a loop at its end.

The stick may now be removed and the butt-end of the arm placed between the halves of the skein.   The skein should appear as in B, fig. 199.

If the skein is formed of hemp or flax and not of horse-hair, the material should be previously soaked in neat's-foot oil.   The oil will preserve the skein and save it from wear and tear ; it will also make the skein into one solid mass, so that when it is twisted up by the winches its strands receive an equal strain.

When a skein is made of fine cord, it will be necessary to wrap this (in forty yard lengths) on a number of large netting needles, such as herring-net makers use.   It would be out of the question to pass and repass fine cord in one length through the winches.

My largest catapult, for instance, required 1,400 yards of cord to make its skein.

When short lengths of fine cord are used, they will have to be knotted together as occasion requires during the process of making the skein.

After the skein is finished and the arm of the catapult has been placed in position therein, the former may be twisted (C, fig. 199).

For this purpose a heavy spanner, 6 ft. long, is necessary.

The eye of the spanner is fitted over the squared spindle (D, fig. 193, p. 280) of one of the winches.   By means of the spanner, three or four men turn one winch slightly.   They then remove the spanner and go round to the opposite side of the catapult and give the other winch a turn.[1]

Numerals may be painted on the large wheels of the winches, so that it may be readily seen if the same number of revolutions are given to each wheel. This is important, as if one winch is turned more than the other the skein will be more tightly twisted on one side of the arm than it is on the other, and a loss of power will ensue.

The winches should be employed to twist up the skein gradually, till it is impossible for three strong men (without the aid of the windlass) to pull the arm back, even a quarter of an inch, from the top cross-bar against which it presses.

Three complete revolutions of the large wheel of each winch should be sufficient to create this amount of pressure.

---

[1] The winches are, of course, always turned in the same direction.

THE SLIP-HOOK ANCIENTLY USED IN LARGE CATAPULTS

FIG. 200, NEXT PAGE

A ring-bolt of wrought iron was secured through the arm of the catapult, just below the part of it which held the stone, figs. 194, 200, pp. 282, 294.

A stout iron slip-hook was then attached to the rope that wound down the arm. The bend of the rope passed through the ring of the slip-hook.

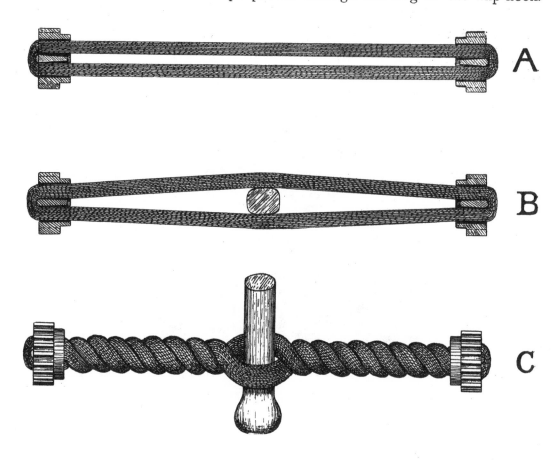

FIG. 199.—THE SKEIN OF CORD IN VARIOUS STAGES.

A. The skein as first wound over the cross-bars of the large wheels of the winches.
B. The skein with the butt-end of the arm placed between its halves.
C. The skein twisted up by the winches.

The point of the slip-hook was hitched inside the eye of the bolt and projected about 1 in. through it, fig. 200, next page.

By pulling the cord attached to the lever of the slip-hook, the point of the latter instantly slipped out of the eye of the bolt and in this way released the arm.

The point of the hook should be short and slightly tapered to its extremity, or it will not easily slip out when required to do so. For the same reason the point of the hook and the inside of the eye of the bolt should be smooth and round.

However great the strain on the slip-hook it will, if properly made, easily effect the release of the arm.

This simple method of releasing the arm of a catapult was far the best as the hook that pulled down the arm was also the means of setting it free.

The slip-hook was able to release the arm at any angle—whether it was fully (as in fig. 200) or only partially wound down.

The trajectory of the weapon was, therefore, controlled by this form of release, as the longer the distance the arm was pulled down the higher the angle at which the projectile was thrown.

FIG. 200.—THE METAL SLIP-HOOK THAT PULLS DOWN THE ARM AND ALSO RELEASES IT.

Its lever or handle is 10 in. long. The point of the hook—which is in the eye of the bolt—is 1 in. thick.

On the other hand, the shorter the distance the arm was drawn back the lower the trajectory of its missile.

If, for instance, a town was being bombarded by a catapult, its arm was wound down to its full extent of 90 degrees; so that the stone it cast might strike the defenders on the ramparts, or else travel high over the defences and fall upon the houses and people inside the walls.

If, however, the besiegers were threatened by a sortie from the gateway of a fortress, the arm of the catapult was set free at a point which was about a quarter less than its full sweep.

Though the force of the missile projected by the catapult was then less than when its arm was fully extended before it was released, the stone travelled low, and bounding along the surface of the ground was more likely to encounter an enemy advancing on horseback or on foot.

In this case the arm of the catapult was wound down to its full extent and could only be set free from this position; hence when the catapult was on level ground the trajectory of its stone did not vary.

To alter the trajectory of the stone thrown by a catapult of this description, the framework of the engine was elevated or depressed, fig. 192, p. 277.

If it was desired to throw a stone at a low trajectory, the after-end of the catapult was raised and wedges were inserted under the ends of its sides.

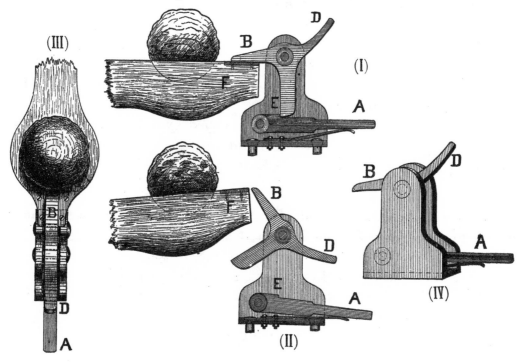

FIG. 201.—THE METAL CATCH FOR A SMALL CATAPULT.

If a high trajectory was required—as when it was wished to drop a stone into a town on an eminence—the front part of the catapult was propped up.[1]

DESCRIPTION OF THE CATCH, FIG. 201

(I) F. The end of the arm of the catapult as held from escaping by the projection B, of the hinged catch D, B.

By knocking down with a mallet the end of the lever A, the leg of the catch (D, B), is freed from the notch in A, at E.

---

[1] Even in the case of the release described in fig. 200, this was also necessary when a fortress was built on ground considerably above the level of the engines attacking it.

(II) The catch (D, B), being then free to swing, the end (F) of the arm of the catapult is instantly released from the projection B, as seen in II, fig. 201.

This figure may also be taken to represent the arm being wound down by the rope and roller.

When the arm is a little lower than shown in II (taking it as coming slowly down and not as flying up), then by lifting the handle (D) of the catch its projection (B) drops over the end (F) of the arm. The leg of the catch at the same time snaps into the notch of the lever A, at E.

In this way the catch is re-set and the arm again secured, as in I fig. 201.

(III) Surface view of the catch holding down the arm.

(IV) Perspective view of the catch.

The iron framework of the catch was bolted to a cross-piece of wood which connected the after-ends of the sides of the catapult.

The roller that wound down the arm was fitted on the front side of this cross-piece, as shown in fig. 198, p. 289.

The rope attached to the roller was hitched by a hook to a ring lashed to the arm, fig. 198, p. 289. When the arm was safely secured by the catch, the rope that pulled it down was unhooked and the catapult was ready for action.

In some catapults, one end of the rope which pulled down the arm was spliced to a cross-bar of metal fixed in the framework of the engine ; its other end being fastened to the winding roller, fig. 198, p. 289.

This arrangement halved the exertion required to pull down the arm and also halved the strain upon the roller, but it doubled the time occupied in winding back the arm.

By using longer levers for turning the roller, the same effect is produced as in the above method and without the loss of time it entailed.

RANGE OF THE CATAPULT

When its skein of cord is tightly twisted, the catapult I have described will pitch a round stone weighing 10 lbs. to a distance of about 350 yards.[1]

Though this is a trivial range when compared with the result obtainable

---

[1] This catapult might easily be fitted with a pair of winches each larger by half than I have given in the plans. This would entail a stronger and slightly longer arm, and also heavier sides to the frame-work of the engine. With these alterations, the catapult would cast a stone weighing 20 lbs.

The stones thrown by catapults do not increase in weight in proportion to the increase in diameter of the skeins of the engines. For example, a catapult with a skein 1 ft. thick will throw a stone three times as heavy as will a catapult with a skein half the size, or 6 in. A skein of 1 ft. in thickness would, however, be double the length of the skein, which was only 6 in. in diameter, as in the former case the framework of the catapult would be much wider than in the latter one.

from a small mortar, it would be a more or less effective one in the days, for example, of the Crusades ; in days when the besiegers camped within a quarter of a mile of the town they were attacking and even conversed with the defenders on its walls.

The great Roman catapult was about twice the size in length and breadth of the one I have given details of.

This immense and powerful machine had an arm of from 10 to 12 ft. long.

An engine of these dimensions—according to the size of its skein—pitched a stone of from 40 to 60 lbs. to a distance of from 350 to 400 yards, the most powerful weapon of the kind being probably able to attain a range of nearly 450 yards.

The velocity of the stone propelled by a catapult was very low as compared with that of a ball from a cannon. It was the ponderous nature of the projectile and not its velocity that did the execution.

A stone of 50 lbs., falling from a short range on battlements and the tops of towers, or among crowded troops and lightly built houses, would be as destructive as a ball of half the weight fired from a cannon at a much longer distance than was possible with a catapult.

The damage to buildings and the slaughter of people must have been terrible, when we consider that 150 to 200 great catapults were often employed at the same time for pounding a city and its defenders, and further, that these engines could be used as freely on the darkest night as by daylight.

Not only were heavy stones thrown among the besieged, their fortifications and their houses, but flaming projectiles were also used which set fire to everything combustible upon which they fell.

Each side of a large catapult was made of two huge logs of wood. The logs were squared and then placed one above the other and bolted together. Winches suitable for twisting a skein of cord such as a 10 to 12 ft. arm required—would necessitate timber of so great a size, that the ancients found it easier to construct the sides of their largest catapults of two longitudinal pieces.

The skein of cord for a catapult with an arm 12 ft. in length, was much larger in proportion to the size of the engine than was the case with a weapon that had a framework of half the dimensions.[1]

---

[1] The catapult with an arm 10 to 12 ft. in length, also cast a stone three times as heavy as that thrown by a weapon half its size.

FIG. 202.—CATAPULT COMPLETED. See Frontispiece for sketches of the engine being worked.

For instance, the catapult I have described has a skein which is 8 in. thick. A catapult double the size of this one in length breadth and height would require a skein about 2 ft. in diameter to work its arm.

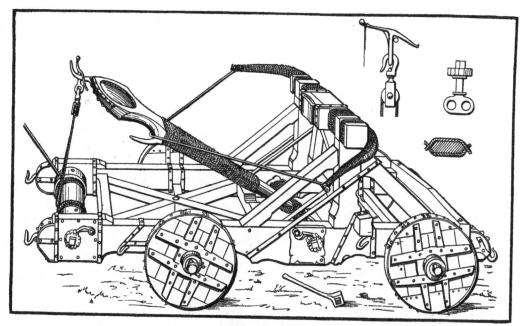

FIG. 203.—A CATAPULT FOR FIELD SERVICE.

Criticism.—This is a good drawing of a catapult, though I cannot find any authority for the addition of the pair of small wooden arms. These are supposed to add to the power of the weapon, but it is doubtful if they would be of the least assistance. In this example, the arm of the engine is not wound down nearly far enough to give sufficient impetus to the stone which will be placed in the hollow of the arm, and the latter is far too weak at its butt-end where it passes through the skein.

The winding roller and the slip-hook for releasing the arm are well shown.

From Viollet-le-Duc.

For practising at a mark on land or water (such as a barrel with a flag erected on its end), an interesting catapult can be made with a framework half the size of the one I have given drawings of.[1]

A small engine of this kind can be worked by a couple of men. It will pitch a cricket ball with great accuracy up to 200 yards and a round smooth stone, the size of a cricket ball, about 250 yards.

[1] The large wheels of the winches should have a bore of $4\frac{1}{2}$ in. and the sides of the framework a height of 15 in. The arm should be 4 ft. long.

FIG. 204.—ROMAN BALISTA WITH ITS BOW-STRING OVER THE CATCH OF THE LOCK AND AN ARROW ON THE STOCK.

CHAPTER LVII

THE BALISTA, ITS CONSTRUCTION AND MANAGEMENT

THE balista projected heavy arrows of a size proportionate to its power and not stones, though it was frequently alluded to by ancient and mediæval writers as a stone-throwing engine.

This mistake arose from the fact that the names balista and catapult were often used indiscriminately in accounts of battles and sieges.

The projectile force of the balista, as in the catapult, was obtained from tightly twisted cordage formed of horse-hair or of the sinews from the necks of large animals, such as horses and oxen.[1]

The construction of the balista resembled two catapults with their arms connected by a thick rope, this rope forming the bow-string of the engine. In appearance the balista was like an immense crossbow, and it doubtless suggested the invention of the crossbow or manubalista carried by the mediæval soldier.

The marked difference between the balista and the crossbow was, that the bow of the balista consisted of two pieces or arms,—each arm being worked by its separate skein of twisted cord,—while the bow of the crossbow was always made in one length.

In the balista, each arm of its bow worked independently. In the crossbow, the arms of the bow acted, of course, as one piece.

The catch which secured the stretched bow-string of the military crossbow was similar to the catch that held the string of the balista ; the lock of the crossbow, as regards its tumbler and long trigger, being closely copied from that of the balista.

The windlass which pulled back the bow-string of the balista was probably the original of the small windlass used for the thick steel bow of the fifteenth century crossbow.

Though mediæval authors sometimes write of this engine as if it had a gigantic bow in one piece, I can find no evidence that such a bow was fitted to the balista.

[1] See footnote, p. 64.

A solid bow of timber would have been of so great a length and thickness to be effective in a balista, that it would, I consider, have been too unwieldy
for active service.   A bow of steel would have been of immense weight to

FIG. 205.—A BALISTA FOR THROWING STONE BALLS.

Criticism.—This engine has six steel arms.   Its mechanism is ingeniously imagined, especially
its lock and the manner of regulating the trajectory of the missile.
      It is possible that an engine of a somewhat similar description was employed by the ancients,
but the huge steel bows shown in the drawing are out of the question, and the absurdity of the one
man bending them collectively by means of a small windlass is evident.

From Ramelli.   Edition of 1588.

transport even had the ancients been able to forge one some 14 to 16 ft. in
length, which is doubtful.
      There was, indeed, no reason why a solid bow, whether of steel or wood,

should be fitted to a balista, for the great power necessary to work the engine was easily procured from skeins of twisted cord; a simpler, much more compact and infinitely lighter method of giving the weapon its projectile force, than fixing to its framework a huge and cumbersome bow in one piece.

Judging from models I have made and from the writings of the ancients, the range of a large balista was from 400 to 450 yards.

The power of a full-sized balista, such as the one shown in fig. 204, p. 300, was immense. The skeins of twisted hair or sinew between which its arms worked, were each about 8 in. thick. The aggregate motive energy of these skeins was, therefore, equal to one skein of 16 in. diameter. Now a catapult with a skein of 16 in. was able to throw a stone weighing from 20 to 30 lbs. from 350 to 400 yards. In the balista there was the same amount of contained force as in the catapult alluded to. In the balista, however, the whole power of the engine was exerted to cast a comparatively light missile, in the form of a javelin of 8 to 10 lbs. weight.[1]

From these considerations it is evident what a terrible weapon the balista must have been when its javelin or huge arrow fell among the ranks of an enemy.

As balistas were much lighter and more portable than catapults, they often represented the field artillery of an army on the march. Catapults of large size were too heavy for transport over rough ground and were essentially siege engines.

Balistas were equally adapted for sieges and for open warfare.

We read of the javelin of a balista piercing through several men, and on another occasion of one of its javelins nailing a man in armour to a tree, p. 271.

This is easily understood when we consider the power of the engine and the distance it was able to throw its steel-tipped projectile.

The balista had an advantage over the catapult in that it was able to cast its missile at a much lower elevation. The balista could be aimed in any direction, and its trajectory could be quickly altered by a couple of men raising or lowering its stock. On the other hand, the ponderous framework of the catapult often had first to be lifted with levers by about a dozen men, and then propped up beneath its ends with wedges to regulate the flight of its stone.

In this way the catapult was fixed for perhaps an entire day so as to batter

---

[1] My largest model of a balista has arms only 2 ft. long and skeins of cordage but 3 in. thick. It will, however, project a javelin in the form of a heavy arrow weighing 2½ lbs., to a range of 300 yards.

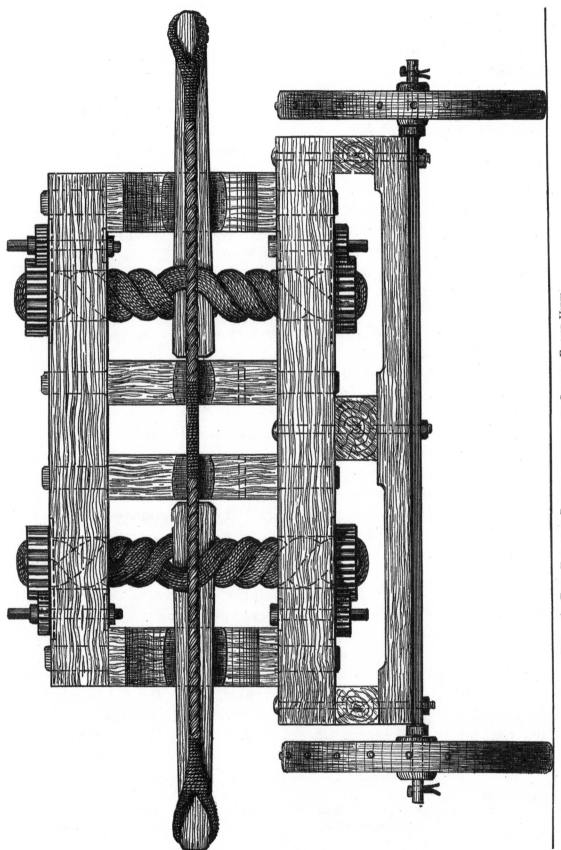

FIG. 206.—FORE END OF A BALISTA WITHOUT ITS STOCK. FRONT VIEW.

down some prominent tower or wall, while the balista could be aimed here and there at the enemy as they showed on the battlements or made sorties.

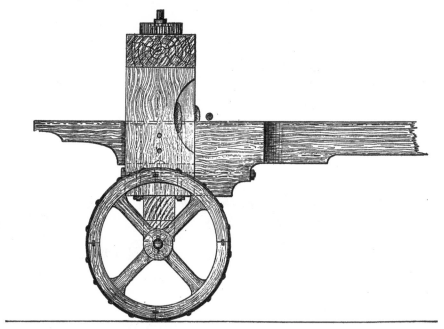

FIG. 207.—FORE-END OF A BALISTA AND ITS STOCK. SIDE VIEW. Scale ½ in. = 1 ft.

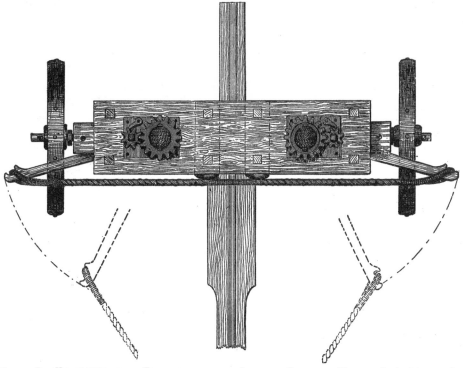

FIG. 208.—FORE-END OF A BALISTA AND ITS STOCK. SURFACE VIEW. Scale ½ in. = 1 ft.

Figs. 204, 206, 207, 208, 209, explain the construction of the balista.   The dimensions of the fore-end of the engine (fig. 206, p. 304) are as follows :

Total length of the oblong frame in which the arms work, 6 ft.

Height of the frame inside, 2 ft.

The four outside timbers of the frame are each 8 in. thick × 18 in. wide.

The two centre pieces of the frame between which the fore-part of the stock is fixed, are each 6 in. thick and 18 in. wide.   These pieces are 7 in. apart.

The skeins of cord are each 8 in. in diameter.

The arms are each 4 ft. long ; 6 in. wide at their butt-ends and 4 in. thick.

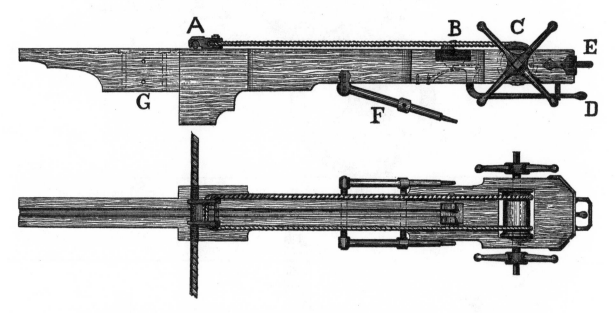

FIG. 209.—SURFACE AND SIDE VIEWS OF THE STOCK OF A BALISTA.   Scale $\frac{1}{2}$ in. = 1 ft.

A. The claws of the windlass hooked over the bow-string preparatory to pulling it back to the catch of the lock.   B. The lock and the catch for holding the stretched bow-string.   C. The windlass.   D. The trigger.   E. The handle for adjusting the stock when aiming.   F. The hinged support which enables the stock to be elevated or depressed to suit the trajectory required. G. The part of the stock that fits between the two uprights in the centre of the framework, see figs. 206, 207, 208.

The length of the stock is 12 ft. ; with a groove along its surface $1\frac{3}{4}$ in. wide and $\frac{1}{2}$ in. deep.

The draw of the bow-string, from a state of rest to the catch of the lock, is $5\frac{1}{2}$ ft.

The narrow part of the stock on which the arrow is laid (fig. 204, p. 300), is 7 in. wide and 8 in. deep.

The winding roller is 7 in. in diameter and 14 in. long.

The large wheels of the four winches that twist the skeins of cordage between which the arms work, have each an opening $8\frac{1}{2}$ in. in diameter.

These winches are identical in design with the one shown in fig. 197, p. 286.

The circular metal tumbler which holds the stretched bow-string and the trigger that releases it, are in every way the same as those described in Chapter XXI, though in this case of course much larger.

The diameter of the tumbler is here $5\frac{1}{2}$ in. and its width 5 in.

The great javelin, or arrow of a balista was from 4 to 6 ft. long, according to the size of the engine. Its shaft, $2\frac{1}{2}$ in. diameter. Its massive steel head weighed from 3 to 5 lbs.

FIG. 210.—THE CLAWS OF THE WINDLASS OF THE BALISTA HOOKED OVER THE BOW-STRING.
Scale $\frac{1}{4}$ full size. See A, fig. 209.

A. The bow-string. B. The rope of the windlass.

The arrow was feathered with two strips of thin wood, horn, or brass, each about 8 in. long. Its butt-end had a cap of metal to save it from being split by the violent contact of the bow-string.

The butt of the arrow was placed between the jaws of the tumbler. It rested against the bow-string as in a crossbow, see fig. 82, p. 128.

The arms of the balista, as they recoiled, struck against small cushions filled with horse-hair. The bow-string was protected from damage in a similar manner, fig. 206, p. 304. The latter was about 2 in. in diameter.

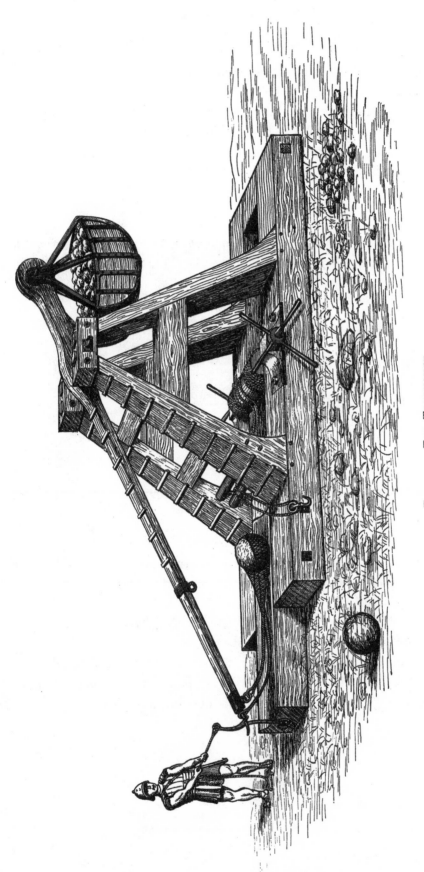

FIG. 211.—THE TREBUCHET.

The arm is fully wound down and the tackle of the windlass is detached from it.   The stone is in the sling and the engine is about to be discharged
by pulling the slip-hook off the end of the arm.   The slip-hook is similar to the one shown in fig. 200, p. 294.
N.B.—A Roman soldier is anachronistically shown in this picture.   The trebuchet was invented after the time of the Romans.

CHAPTER LVIII

THE TREBUCHET

This engine was of much more recent invention than either the catapult or the balista of the Greeks and Romans. It is said to have been introduced into siege operations by the French in the twelfth century. On the other hand, the catapult and the balista were in use before the Christian Era. Egidio Colonna gives a fairly accurate description of the trebuchet, and writes of it about 1280 as though it were the most effective siege weapon of his time.

The projectile force of this weapon was obtained from the terrestrial gravitation of a heavy weight, and not from twisted cordage as in the catapult and balista.

From about the middle of the thirteenth century, the trebuchet in great measure superseded the catapult. This preference for the trebuchet was due to the fact that it was able to cast stones of 300 lbs. and more in weight, or five or six times as heavy as those which the largest catapults could project.

The stones of 50 to 60 lbs. thrown by siege catapults would no doubt destroy towers and battlements, as the result of the constant and concentrated bombardment of many engines. One huge stone of 300 lbs., as slung from a trebuchet, would however shake the strongest defensive masonry and easily break through the upper parts of the walls of a fortress.

The trebuchet was essentially an engine for destroying the defences of a fortification, so that it might be entered by means of scaling ladders or in other ways.

From experiments with models of good size and from other sources, I find that the largest trebuchets—those with arms of about 50 ft. in length and counterpoises of about 20,000 lbs.—were capable of slinging a stone 300 lbs. in weight to a distance of 300 yards, a range of 350 yards being in my opinion more than these engines were able to attain.[1]

---

[1] Egidio Colonna tells us that the trebuchet was sometimes made without a counterpoise, and that in such a case the arm of the engine was worked by a number of men pulling together instead of by a heavy weight. I cannot believe this, as however many men pulled at the arm of a trebuchet they could not apply nearly the force that would be conveyed by the terrestrial gravitation of a heavy weight.

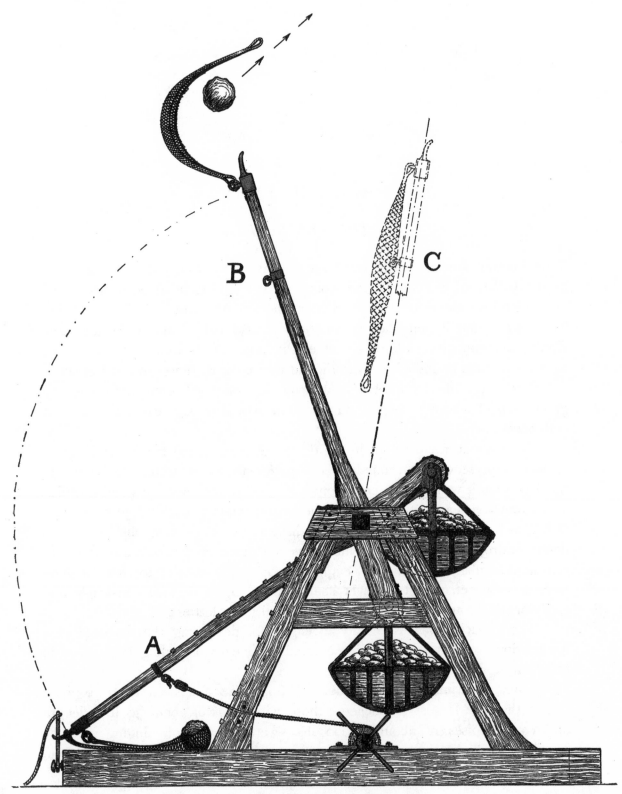

FIG. 212.—THE ACTION OF THE TREBUCHET.

A. The arm pulled down and secured by the slip-hook previous to unhooking the rope of the windlass.   B. The arm released
from the slip-hook and casting the stone out of its sling.   C. The arm at the end of its upward sweep.

The trebuchet made by order of Napoleon III., and described in his ' Etudes sur l'artillerie,' had an arm 33 ft. in length with a counterpoise of 10,000 lbs.

FIG. 213.—A TREBUCHET WITH ITS ARM BEING WOUND DOWN.

Criticism.—Here we have a trebuchet with an arm at least 60 ft. in length.

An engine of such immense size as this would require a score of men at its windlass instead of a couple.

The heavy stone was placed in the great sling of thick netting which is suspended to the end of the arm.

The sling was identical in its action with the one given in fig. 212. See also fig. 181, p. 253.

From Viollet-le-Duc.

weight to work it. This machine projected a 50 lb. cannon-ball 200 yards, but was so lightly constructed that its full power could not be safely applied.

In a book on 'Experimental Philosophy,' by J. T. Desaguliers, 1734, a curious and interesting old work on mechanical effects, the author gives

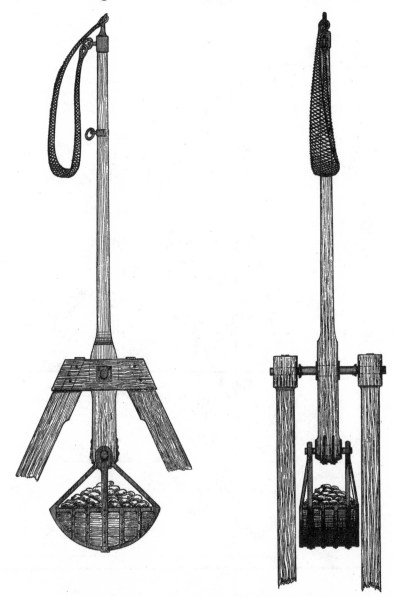

FIG. 214.—THE ARM OF THE TREBUCHET AND ITS COUNTERPOISE.
FRONT AND SIDE VIEW.

a detailed calculation of the power of a trebuchet, together with plans of the engine as constructed from the writings of Vitruvius.

These drawings are, however, inaccurate, and though Desaguliers' conclusions are exact, he only allows the trebuchet a counterpoise of 2,000 lbs., which would be far too light a weight to be of any service in an engine of the kind.

The trebuchet is sometimes depicted in mediæval books with an arm like

FIG. 215.—TREBUCHETS THROWING BARRELS FILLED WITH EARTH INTO THE DITCH OUTSIDE A FORTRESS, SO AS TO ENABLE THE BESIEGERS TO PASS OVER IT AND APPLY THEIR SCALING LADDERS TO THE WALLS.

Criticism.—A very elaborate and fanciful drawing. The counterpoise of the nearer engine could not swing back between the uprights, and it and the other engine would each require at least six men to work the windlasses.

The barrels would not be projected 30 yards and the men working the engines would be slain by the archers on the battlements of the besieged fortress.

From Ramelli.   Edition of 1588.

that of a catapult (*i.e.* with a hollow in the end of the arm in which to rest the stone as in fig. 190, p. 273), and without a sling, but this is incorrect.

The trebuchet always had a sling in which to place its missile.

The sling doubled the power of the engine and caused it to throw its projectile twice as far as it would have been able to do without it.

It was the length of the arm, when suitably weighted with its counterpoise, which combined with its sling gave power to the trebuchet. Its arm, when released, swung round with a long easy sweep and with nothing approaching the velocity of the much shorter arm of the catapult.

The weight of the projectile cast by a trebuchet was governed by the weight of its counterpoise. Provided the engine was of sufficient strength and could be manipulated, there was scarce a limit to its power. Numerous references are to be found in mediæval authors to the practice of throwing dead horses into a besieged town with a view to causing a pestilence therein, and there can be no doubt that trebuchets were employed for this purpose. As a small horse weighs about 10 cwt., we can form some idea of the size of the rocks and balls of stone that trebuchets were capable of slinging.

When we consider that a trebuchet was able to throw a horse over the walls of a town we can credit the statement of Stella[1] who writes 'that the Genoese armament sent against Cyprus in 1373 had among other great engines one which cast stones of 12 cwt.'

Villard de Honnecourt[2] describes a trebuchet that had a counterpoise of sand the frame of which was 12 ft. long, 8 ft. broad, and 12 ft. deep. That such machines were of vast size will readily be understood. For instance, twenty-four engines taken by Louis IX at the evacuation of Damietta in 1249, afforded timber for stockading his entire camp ;[3] a trebuchet used at the capture of Acre by the Infidels in 1291, formed a load for an hundred carts ;[4] a great engine that cumbered the tower of St. Paul at Orleans and which was dismantled previous to the celebrated defence of the town against the English in 1428-9, furnished twenty-six cart loads of timber.[5]

All kinds of articles besides horses, men, stones and bombs were at times

---

[1] Stella G. Flourished at the end of the fourteenth century and beginning of fifteenth. He wrote *The Annals of Genoa* from 1298-1409. Muratori includes the writings of Stella in his great work, *Rerum Italicarum Scriptores*, 25 vols. 1723-38.

[2] Villard de Honnecourt, an engineer of the thirteenth century. His album translated and edited by R. Willis, M.A., 1859.

[3] Jean Sire de Joinville. He went with St. Louis to Damietta. His memoirs, written in 1309, published by F. Michel, 1858.

[4] Abulfeda, 1273-1331. Arab soldier and historian, wrote *Annals of the Moslems*. Published by Hafnire, 1789-94. Abulfeda was himself in charge of one of the hundred carts.

[5] From an old history of the siege (in manuscript) found in the town hall of Orleans and printed by Saturnin Holot, a bookseller of the city, 1576.

thrown from trebuchets. Vassaf[1] records 'that when the garrison of Delhi refused to open the gates to Ala'uddin Khilji in 1296, he loaded his mangonels with bags of gold and shot them into the fortress, a measure which put an end to the opposition.'

Figs. 211, 212, 214, explain the construction and working of a trebuchet.

[1] Persian historian, lived at end of thirteenth and beginning of fourteenth century. The preface to his history is dated 1288, and the latter is carried down to 1312.

CHAPTER LIX

THE SPRING ENGINE

I CALL this machine ' the Spring engine ' as I am ignorant of its distinctive name.[1] It is possible it may have been the ' Espringale,' ' Espringold,' or ' Springald ' but there is no evidence to show that this was the case, or, indeed, what it was called.

FIG. 216.—A SPRING ENGINE.

Criticism.—This engine is well depicted, together with the slip-hook which releases the arm or spring, when the latter is drawn back far enough to strike the butt-end of the javelin with sufficient force to project it.

From Viollet-le-Duc.

The spring engine, as portrayed in mediæval works, appears to have been of simple construction though it was no doubt an effective weapon. All the mediæval drawings of it with which I am acquainted are, however, very crude,

[1] Of all the siege engines of the ancients this was certainly the most primitive as regards its propelling mechanism.

FIG. 217.—A SPRING ENGINE.

Criticism.—A drawing of very mediæval character. It shows, however, the principle of the weapon.

The grooved block on which the arrow was laid was raised or lowered by the prop beneath it. The upper end of the prop was hinged to the block which supported the arrow. Its lower end rested on the notches of the upright according to the trajectory required. The arm or spring was pulled back by a windlass and tackle, and on being released violently struck the butt-end of the arrow and projected it forward.

From Vegetius. Edition of 1607.

and though they give a general idea of the machine their details of its mechanism are often confusing.

The motive principle of the spring engine was precisely as if you placed a wooden match $\frac{1}{2}$ in. over the edge of a table, and then struck the projecting part of the match with the blade of a steel knife, first by bending the blade of the knife back with the fingers and then suddenly releasing it.

The match may be taken to represent the javelin and the blade of the knife the spring that propels it.

Fig. 218.—A Spring Engine.

Criticism.—A complicated and fanciful drawing of a spring engine. The arrows would be broken off at their butt-ends instead of being projected forward.

From Vegetius. Edition of 1607.

The arm, or spring of this engine was of such great strength that it could not possibly be bent without the aid of a windlass. It is probable that the arm consisted of many thin flat laths of elastic wood, glued and bound together. A steel spring would have been too stiff, I consider, for propelling the javelin.

The arm was secured in a strong frame, and was drawn back by a rope and winch till it was sufficiently bent to serve its purpose. It was then set free by a slip-hook and instantly struck the projecting butt-end of the javelin a violent blow.

The arm was bent and then retained in this condition by a slip-hook similar to the one shown in fig. 216, p. 316. The arrow rested on a block of wood with a groove along its centre as in the stock of a crossbow. The trajectory of the arrow was arranged by means of an adjustable prop. This was hinged to the under-side of the piece of wood on which the missile was placed, fig. 217, p. 317.

That the spring engine was a serviceable one when of large size is beyond question, though its range and power are conjectural. With a model I have discharged a crossbow bolt 160 yards, the wooden spring of the model being 5 ft. in length and formed of eight strips of ash each 3 in. broad and $\frac{1}{4}$ in. thick.

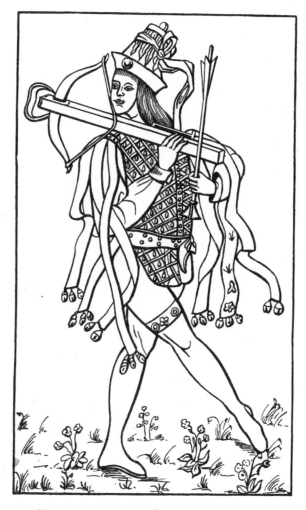

FROM 'ANCIENT TAPESTRIES.' A. JUBINAL. PARIS, 1838.

VALE.

INDEX

The numerals in parentheses refer to periods and dates

APPENDIX

CONTENTS OF THE APPENDIX

THE CATAPULT AND BALISTA

THE TURKISH COMPOSITE BOW

SINCE my recent work on the crossbow and ancient projectile weapons was issued,[1] I have obtained additional information concerning the catapult and balista of the Greeks and Romans. I now, therefore, print, in the form of an Appendix, a revised account of the construction of these two engines. Their history and effects in warfare I have already dealt with.

I also append a treatise fully describing the structure, power and management of that remarkable weapon the Turkish composite bow, which I only cursorily alluded to in my book on the crossbow etc.

R. P. G.

THIRKLEBY PARK,
        THIRSK :
            Jan. 1907.

---

[1] *The Crossbow, Mediæval and Modern, Military and Sporting: its Construction, History, and Management. With a Treatise on the Balista and Catapult of the Ancients.* 220 illustrations. Messrs. Longmans & Co., 39 Paternoster Row, London.

INTRODUCTORY NOTES ON ANCIENT
PROJECTILE ENGINES

OF ancient Greek authors who have left us accounts of these engines, Heron (284–221 B.C.) and Philo (about 200 B.C.) are the most trustworthy.

Both these mechanicians give plans and dimensions with an accuracy that enables us to reconstruct the machines, if not with exactitude at any rate with sufficient correctness for practical application.

Though in the books of Athenæus, Biton, Apollodorus, Diodorus, Procopius, Polybius and Josephus we find incomplete descriptions, these authors, especially Josephus, frequently allude to the effects of the engines in warfare; and scanty as is the knowledge they impart, it is useful and explanatory when read in conjunction with the writings of Heron and Philo.

Among the Roman historians and military engineers, Vitruvius and Ammianus are the best authorities.

Vitruvius copied his descriptions from the Greek writers, which shows us that the Romans adopted the engines from the Greeks.

Of all the old authors who have described the engines, we have but copies of the original writings. It is, therefore, natural that we should come across many phrases and drawings which are evidently incorrect, as a result of repeated transcription, and which we know to be at fault though we cannot actually prove them to be so.

With few exceptions, all the authors named simply present us with their own ideas when they are in doubt respecting the mechanical details and performances of the engines they wish to describe.

All such spurious information is, of course, more detrimental than helpful to our elucidation of their construction and capabilities.

It frequently happens that in a mediæval picture of one of these machines some important mechanical detail is omitted, or, from the difficulty of portraying it correctly, is purposely concealed by figures of soldiers, an omission that may be supplied by reference to other representations of the same weapon.

It is, indeed, impossible to find a complete working plan of any one of these old weapons, a perfect design being only obtainable by consulting many ancient authorities, and, it may be said, piecing together the details of construction they individually give.

We have no direct evidence as to when the engines for throwing projectiles were invented.

It does not appear that King Shalmaneser II. of Assyria (859–825 B.C.) had any, for none are depicted on the bronze doors of the palace of Balâwat, now in the British Museum, on which his campaigns are represented, though his other weapons of attack and defence are clearly shown.

The earliest allusion is the one in the Bible, where we read of Uzziah, who reigned from B.C. 808–9 to B.C. 756–7. 'Uzziah made in Jerusalem engines invented by cunning men, to be on the towers and upon the bulwarks, to shoot arrows and great stones withal.' (2 Chronicles xxvi. 15.)

Diodorus tells us that the engines were first seen about 400 B.C., and that when Dionysius of Syracuse organised his great expedition against the Carthaginians (397 B.C.) there was a genius among the experts collected from all over the world, and that this man designed the engines that cast stones and javelins.

From the reign of Dionysius and for many subsequent centuries, or till near the close of the fourteenth, projectile-throwing engines are constantly mentioned by military historians.

But it was not till the reign of Philip of Macedon (360–336 B.C.) and that of his son Alexander the Great (336–323 B.C.) that their improvement was carefully attended to and their value in warfare fully recognised.

As before stated, the Romans adopted the engines from the Greeks.

Vitruvius and other historians tell us this, and even copy their descriptions of them from the Greek authors, though too often with palpable inaccuracy.

To ascertain the power and mechanism of these ancient engines a very close study of all the old authors who wrote about them is essential, with a view to extracting here and there useful facts amid what are generally verbose and confused references.

There is no doubt that the engines made and used by the Romans after their conquest of Greece (B.C. 146), in the course of two or three centuries became inferior to the original machines previously constructed by the Greek artificers.

Their efficiency chiefly suffered because the art of manufacturing their important parts was gradually neglected and allowed to become lost.

For instance, how to make the skein of sinew that bestowed the very life and existence on every projectile-casting engine of the ancients.

The tendons of which the sinew was composed, the animals from which it was taken, and the manner in which it was prepared, we can never learn now.

Every kind of sinew, or hair or rope, with which I have experimented, either breaks or loses its elasticity in a comparatively short time, if great pressure is applied. It has then to be renewed at no small outlay of expense and trouble. Rope skeins, with which we are obliged to fit our models, cannot possibly equal in strength, and above all in elasticity, skeins of animal sinew or even of hair.

The formation of the arm or arms of an engine, whether it is a catapult with its single upright arm or a balista with its pair of lateral ones, is another difficulty which cannot now be overcome, for we have no idea how these arms were made to sustain the great strain they had to endure.

We know that the arm of a large engine was composed of several spars of wood and lengths of thick sinew fitted longitudinally, and then bound round with broad strips of raw hide which would afterwards set nearly as hard and tight as a sheath of metal.

We know this, but we do not know the secret of making a light and flexible arm of sufficient strength to bear such a strain as was formerly applied to it in a catapult or a balista.

Certainly, by shaping an arm of great thickness we can produce one that will not fracture, but substance implies weight, and undue weight prevents the arm from acting with the speed requisite to cast its projectile with good effect.

A heavy and ponderous arm of solid wood cannot, of course, rival in lightness and effectiveness a composite one of wood, sinew and hide.

The former is necessarily inert and slow in its action of slinging a stone, while the latter would, in comparison, be as quick and lively as a steel spring.

When the art of producing the perfected machines of the Greeks was lost, they were replaced by less effective contrivances.

If the knowledge of constructing the great catapult of the ancients in its original perfection had been retained, such a clumsy engine as the mediæval trebuchet would never have gained popularity. The trebuchet derived its power from the gravity of an immense weight at one end of its pivoted arm tipping up the other end, to which a sling was attached for throwing a stone.

As regards range, there could be no comparison between the efficiency of a trebuchet, however large, as worked merely by a counterpoise, and that of an engine deriving its power from the elasticity of an immense coil of tightly twisted sinew.

It is certain that if the latter kind of engine had survived in its perfect state the introduction of cannon would have been considerably delayed, for the effects in warfare of the early cannon were for a long period decidedly inferior to those of the best projectile engines of the ancients.

Notwithstanding many difficulties, I have succeeded in reconstructing, though of course on a considerably smaller scale, the chief projectile-throwing engines of the ancients, and with a success that enables them to compare favourably, as regards their range, with the Greek and Roman weapons they represent.

Still, my engines are by no means perfect in their mechanism, and are, besides, always liable to give way under the strain of working.

One reason of this is that all modern engines of the kind require to be worked to their utmost capacity, *i.e.* to the verge of their breaking point, to obtain from them results that at all equal those of their prototypes.

A marked difference between the ancient engines and their modern imitations, however excellent the latter may be, is, that the former did their work easily, and well within their strength, and thus without any excessive strain which might cause their collapse after a short length of service.[1]

The oft-disputed question as to the distance to which catapults and balistas shot their projectiles can be solved with approximate accuracy by comparing their performances—as given by ancient military writers—with the results obtainable from modern reproductions.

While treating of this matter we should carefully consider the position and surroundings of the engines when engaged in a siege, and especially the work for which they were designed.

As an example, archers, with the advantage of being stationed on high towers and battlements, would be well able to shoot arrows from 270 to 280 yards. For this reason it was necessary for the safe manipulation of the attacking engines that they should be placed at about 300 yards from the outer walls of any fortress they were assailing.

As a catapult or a balista was required not only to cast its missile among the soldiers on the ramparts of a fortified place, but also to send it clear over the walls amid the houses and people within the defences, it is evident that the

---

[1] Again, though my largest catapult will throw a stone to a great distance it cannot throw one of nearly the weight it should be able to do, considering the size of its frame, skein of cord and mechanism. In this respect it is decidedly inferior to the ancient engine.

engines must have had a range of from 400 to 500 yards, or more, to be as serviceable and destructive as they undoubtedly were.

Josephus tells us that at the siege of Jerusalem, A.D. 70 ('Wars of the Jews,' Book V. Chapter VI.), stones weighing a talent (57¾ lbs. avoirdupois) were thrown by the catapults to a distance of two or more 'stades.'

This statement may be taken as trustworthy, for Josephus relates what he personally witnessed and his comments are those of a commander of high rank and intelligence.

Two or more 'stades,' or let us say 2 to 2¼ 'stades,' represent 400 to 450 yards. Remarkable and conclusive testimony confirming the truth of what we read in Josephus is the fact that my largest catapult—though doubtless much smaller and less powerful than those referred to by the historian—throws a stone ball of 8 lbs. in weight to a range of from 450 to nearly 500 yards.

It is easy to realise that the ancients, with their great and perfect engines fitted with skeins of sinew, could cast a far heavier stone than one of 8 lbs., and to a longer distance than 500 yards.

Agesistratus,[1] a Greek writer who flourished B.C. 200, and who wrote a treatise on making arms for war, estimated that some of the engines shot from 3½ to 4 'stades' (700 to 800 yards).

Though such a very long flight as this appears almost incredible, I can adduce no sound reason for doubting its possibility. From recent experiments I am confident I could now build an engine of a size and power to accomplish such a feat if light missiles were used, and if its cost were not a consideration.

---

[1] The writings of Agesistratus are non-extant but are quoted by Athenæus.

FIG. 1.—SKETCH PLAN OF A CATAPULT FOR SLINGING STONES, ITS ARM BEING PARTLY WOUND DOWN.

Approximate scale : ½ in. = 1 ft.

THE CATAPULT (WITH A SLING)

THE mediæval catapult was usually fitted with an arm that had a hollow or cup at its upper end in which was placed the stone it projected.[1] I find, however, that the original and more perfect form of this engine, as employed by the Greeks and ancient Romans, had a sling, made of rope and leather, attached to its arm.[2] (Fig. 1, opposite page.)

The addition of a sling to the arm of a catapult increases its power by at least a third. For example, the catapult described in Chapters LV., LVI., of my book,[3] will throw a round stone 8 lbs. in weight, from 350 to 360 yards, but the same engine with the advantage of a sling to its arm will cast the 8-lb. stone from 450 to 460 yards, and when its skein is twisted to its limit of tension to nearly 500 yards.

If the upper end of the arm of a catapult is shaped into a cup to receive the stone, as shown in figs. 187, 192, pp. 267, 277 of 'The Crossbow,' the arm is, of necessity, large and heavy at this part.

If, on the other hand, the arm is equipped with a sling, as shown in fig. 1, opposite page, it can be tapered from its butt-end upwards, and is then much lighter and recoils with far more speed than an arm that has an enlarged extremity for holding its missile.

When the arm is fitted with a sling, it is practically lengthened by as much as the length of the sling attached to it, and this, too, without any appreciable increase in its weight.

The longer the arm of a catapult, the longer is its sweep through the air, and thus the farther will it cast its projectile, provided it is not of undue weight.

---

[1] See *The Crossbow, etc.*, Chapters LV., LVI., illustrations 193 to 202.

[2] In mediæval times catapults which had not slings cast great stones, but only to a short distance in comparison with the earlier weapons of the same kind that were equipped with slings. I can find no allusions or pictures to show that during this period any engine was used with a sling except the trebuchet, a post-Roman invention. All evidence goes to prove that the secret of making the skein and other important parts of a catapult was in a great measure lost within a couple of centuries after the Romans copied the weapon from their conquered enemies the Greeks, with the result that the trebuchet was introduced for throwing stones.

The catapult was gradually superseded as the art of its construction was neglected, and its efficiency in sieges was therefrom decreased.

The catapults of the fifth and sixth centuries were very inferior to those described by Josephus as being used at the sieges of Jerusalem and Jotapata (A.D. 70, A.D. 67).

[3] *The Crossbow, etc.*

The difference in this respect is as between the range of a short sling and that of a long one, when both are used by a school-boy for slinging pebbles.

The increase of power conferred by the addition of a sling to the arm of a catapult is surprising.

A small model I constructed for throwing a stone ball, 1 lb. in weight, will attain a distance of 200 yards when used with an arm that has a cup for holding the ball, though when a sling is fitted to the arm the range of the engine is at once increased to 300 yards.

The only historian who distinctly tells us that the catapult of the Greeks and Romans had a sling to its arm, is Ammianus Marcellinus. This author flourished about 380 A.D., and a closer study of his writings, and of those of his contemporaries, led me to carry out experiments with catapults and balistas which I had not contemplated when my work dealing with the projectile engines of the ancients was published.

Ammianus writes of the catapult[1]:

'In the middle of the ropes[2] rises a wooden arm like a chariot pole . . . to the top of the arm hangs a sling . . . when battle is commenced a round stone is set in the sling . . . four soldiers on each side of the engine wind the arm down till it is almost level with the ground . . . when the arm is set free it springs up and hurls forth from its sling the stone, which is certain to crush whatever it strikes. This engine was formerly called the " scorpion," because it has its sting erect,[3] but later ages have given it the name of Onager, or wild ass, for when wild asses are chased they kick the stones behind them.'

FIG. 2.—CATAPULT (WITH A SLING), SEE OPPOSITE PAGE.

A. The arm at rest, ready to be wound down by the rope attached to it and also to the wooden roller of the windlass. The stone may be seen in the sling.

The upper end of the pulley rope is hitched by a metal slip-hook (fig. 1, p. 10) to a ring-bolt secured to the arm just below the sling.

B. The position of the arm when fully wound down by means of the windlass and rope. See also EE, fig. 3, p. 14.

C. The position of the arm at the moment the stone D leaves the sling, which it does at an angle of about 45 degrees.

---

[1] *Roman History*, Book XXIII., Chapter IV.

[2] *i.e.* in the middle of the twisted skein formed of ropes of sinew or hair.

[3] The upright and tapering arm of a catapult, with the iron pin on its top for the loop of the sling, is here fancifully likened to the erected tail of an angry scorpion with its sting protruding.

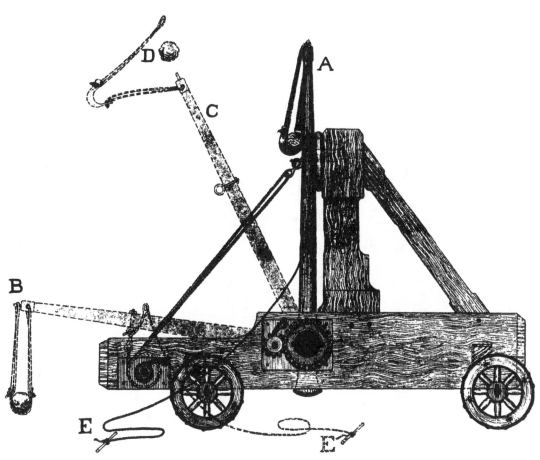

Fig. 2.—Catapult (with a Sling). Side view of frame and mechanism.

Scale : $\frac{1}{8}$ in. = 1 ft.

E.   By pulling the cord E the arm B is at once released from the slip-hook and, taking an upward sweep of 90 degrees, returns to its original position at A.

THE SLING (OPEN).

[F.   Its fixed end which passes through a hole near the top of the arm.

G.   The leather pocket for the stone.

H.   The loop which is hitched over the iron pin at the top of the arm when the stone is in position in the sling, as shown at A and B, fig. 2, p. 13.]

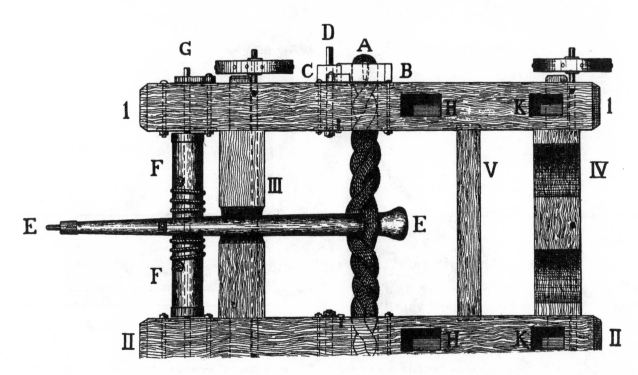

FIG. 3.—CATAPULT (WITH A SLING). Surface view of frame and mechanism. Scale : ½ in. = 1 ft. The arm EE is here shown wound down to its full extent. (Compare with B, fig. 2, p. 13.)

I.   I.   } The side-pieces.
II.   II.

III.   IV.   The large cross-pieces.

       V.   The small cross-piece.

The ends of the cross-piece beams are stepped into the side-pieces.

AA. The skein of twisted cord.

BB. The large winding wheels. The skein is stretched between these wheels, its ends passing through the sides of the frame, and then through the wheels and over their cross-bars. (Fig. 6, p. 17.)

By turning with a long spanner (fig. 1, p. 10) the squared ends of the spindles DD, the pinion wheels CC rotate the large wheels BB and cause the latter to twist the skein AA, between the halves of which the arm EE is placed.

FF. The wooden roller which winds down the arm EE. (Fig. 1, p. 10.)

The roller is revolved by four men (two on each side of the engine) who fit long spanners on the squared ends of the iron spindle GG.

This spindle passes through the centre of the roller and through the sides of the frame.

The small cogged wheels, with their checks, which are fitted to the ends of the spindle GG, prevent the roller from reversing as the arm is being wound down. (Fig. 1, p. 10.)

HH. The hollows in the sides of the frame which receive the lower tenons of the two uprights. Between the tops of these uprights the cross-beam is fixed against which the arm of the catapult strikes when it is released. (Fig. 1, p. 10.)

KK. The hollows for the lower tenons of the two sloping supports which prevent the uprights, and the cross-beam between them, from giving way when the arm recoils. (Fig. 1, p. 10.)

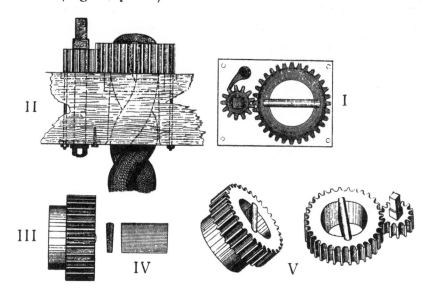

FIG. 4.—ONE OF THE PAIR OF WINCHES OF A CATAPULT. Scale : $\frac{1}{10}$ in. = 1 in.

I. Surface view of one of the winches and of the thick iron plate in which the socket of the large winding wheel of the winch revolves.

II.   View of a winch (from above) as fitted into one of the sides of the frame of the catapult.   One end of the twisted skein may be seen turned round the cross-bar of the large wheel.

III.   Side view of the large wheel of a winch.

IV.   The cross-bar of one of the large wheels.   These pieces fit like wedges into tapering slots cut down the barrels, or inside surfaces, of their respective wheels.

V.   Perspective view of the wheels of a winch.

The winches are the vital parts of the catapult, as they generate its projectile power.

They are employed to twist tightly the skein of cord between which the butt-end of the arm of the engine is placed.

The cord composing the skein is stretched to and fro across and through the sides of the catapult, and alternately through the insides of the large wheels and over their cross-bars ; as shown in fig. 3, p. 14.

FIG. 5.—THE IRON SLIP-HOOK.

FIG. 5.

This simple contrivance not only pulled down the arm of a catapult but was also the means of setting it free. However great the strain on the slip-hook, it will, if properly shaped, easily effect the release of the arm.

The trajectory of the missile can be regulated by this form of release, as the longer the distance the arm is pulled down the higher the angle at which the projectile is thrown.

On the other hand, the shorter the distance the arm is drawn back, the lower the trajectory of its missile.

The slip-hook will release the arm of the engine at any moment, whether it is fully or only partially wound down by the windlass.

The slip-hook of the large catapult shown in fig. 1, p. 10, has a handle, *i.e.* lever, 10 inches long, the point of the hook, which passes through the eye-bolt secured to the arm, being 1 in. in diameter.

FIG. 6.—THE SKEIN OF CORD, SEE OPPOSITE PAGE.

A.   The skein as first wound over the cross-bars of the large wheels (shown in section) of the winches.

B.   The skein with the butt-end of the arm (shown in section) placed between its halves.

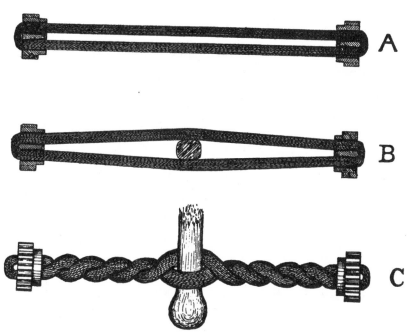

FIG. 6.—THE SKEIN OF CORD.

C. The skein as it appears when tightly twisted up by the winches. Compare with AA, fig. 3, p. 14.

Cord of Italian hemp, about $\frac{1}{4}$ in. thick, is excellent for small catapults. For large ones, horsehair rope, $\frac{1}{2}$ in. thick, is the best and most elastic. Whatever is used, the material of the skein must be thoroughly soaked in neat's-foot oil for some days previously, or it is sure to fray and cut under the friction of being very tightly twisted. Oil will also preserve the skein from damp and decay for many years.

HOW TO WORK THE CATAPULT

There is little to write under this heading; as the plans, details of construction and illustrations will, I trust, elucidate its management.

The skein should never remain in a tightly twisted condition, but should be untwisted when the engine is not in use.

Previous to using the catapult its winches should be turned with the long spanner, fig. 1, p. 10, first the winch on one side of the engine and then the one on the other side of it, and each to exactly the same amount.

Small numerals painted on the surfaces of the large wheels, near their edges, will show how much they have been revolved; in this way their rotation can be easily arranged to correspond.

As the skein of cord is being twisted by the very powerful winches, the arm will gradually press with increasing force against the cross-beam between the uprights.   The arm should be so tightly pressed against the fender, or cushion of straw, attached to the centre of this beam, that it cannot be pulled back the least distance by hand.

If the skein of my largest catapult is fully tightened up by the winches, three strong men are unable to draw the arm back with a rope even an inch from the cross-beam, though the windlass has to pull it down from six to seven feet when the engine is made ready for action.

When the skein is as tight as it should be, attach the slip-hook to the ring-bolt in the arm and place the stone in the sling suspended from the top of the arm.

The arm can now be drawn down by means of long spanners fitted to the windlass.   Directly the arm is as low as it should be, or as is desired, it should be instantly released by pulling the cord fastened to the lever of the slip-hook.

The least delay in doing this, and the resulting continuation of the immense strain on the arm, may cause it to fracture when it would not otherwise have done so.

The plans I have given are those of my largest engine, which, ponderous as it seems—(it weighs two tons)—is, however, less than half the size of the catapult used by the ancients for throwing stones of from forty to fifty pounds in weight.

As the plans are accurately drawn to scale, the engine can easily be reproduced in a smaller size.

An interesting model can be constructed that has an arm 3 feet in length, and a skein of cord about 4 inches in diameter.   It can be worked by one man and will throw a stone, the size of an orange, to a range of 300 yards.

The sling, when suspended with the stone in position, should be one third the length of the arm, as shown in fig. 2, p. 13.

If the sling is shortened, the ball will be thrown at a high elevation.   If the sling is lengthened, the ball will travel at a lower angle and with much more velocity.

THE BALISTA

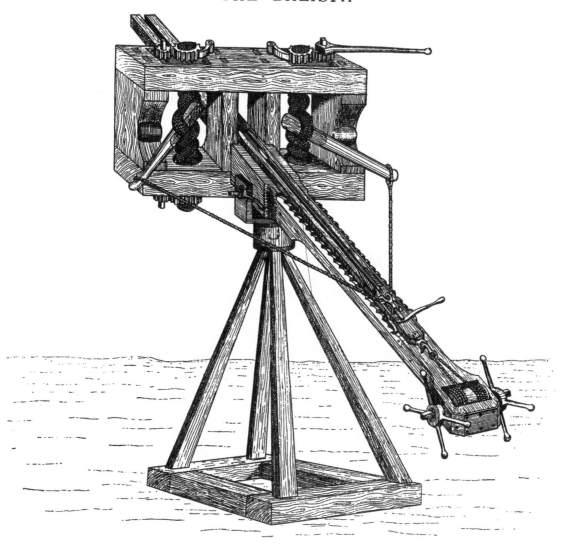

FIG. 7.—BALISTA FOR DISCHARGING HEAVY ARROWS OR JAVELINS.
Approximate scale : $\frac{1}{2}$ in. = 1 ft.

THIS engine is here shown ready for discharge with its bow-string drawn to its full extent by the windlass.

The heavy iron-tipped arrow rests in the shallow wooden trough which travels along the stock.

The trough has a strip of wood, in the form of a keel, fixed beneath it. This keel travels to or fro in a dovetailed slot cut along the upper surface of the stock for the greater part of its length.   (F, fig. 8, p. 21.)

The arrow is laid in the trough before the bow-string is stretched.   (A, B, fig. 8, p. 21.)

The balista is made ready for use by turning the windlass.   The windlass pulls back the sliding trough, and the arrow resting in it, along the stock of the engine, till the bow-string is at its proper tension for discharging the projectile. (Fig. 7, p. 19.)

As the trough and the arrow are drawn back together, the arrow can be safely laid in position before the engine is prepared for action.

The catch for holding the bow-string, and the trigger for releasing it, are fixed to the solid after-end of the wooden trough.   (Fig. 8, p. 21.)

The two ratchets at the sides of the after-end of the trough travel over and engage, as they pass along, the metal cogs fixed on either side of the stock. (Fig. 8, p. 21.) [1]

By this arrangement the trough can be securely retained, in transit, at any point between the one it started from and the one it attains when drawn back to its full extent by the windlass.

As the lock and trigger of the balista are fixed to the after-end of the sliding trough (G, fig. 8, p. 21), it will be realised that the arrow could be discharged at any moment required in warfare, whether the bow-string was fully or only partially stretched.

In this respect the balista differed from the crossbow, which it somewhat resembled, as in a crossbow the bow-string cannot be set free by the trigger at an intermediate point, but only when it is drawn to the lock of the weapon.

It will be seen that the balista derives its power from two arms; each with its separate skein of cord and pair of winches.

These parts of the balista are the same in their action and mechanism as those of the catapult.

FIG. 8 (OPPOSITE PAGE).—THE MECHANISM OF THE STOCK OF AN ARROW-THROWING BALISTA.

A.   Side view of the stock, with the arrow laid in the sliding trough before the bow-string is stretched.

B.   Surface view of the stock, with the arrow laid in the sliding trough before the bow-string is stretched.

C.   Section of the fore-end of the stock, and of the trough which slides in and along it.

---

[1] When the bow-string has been released and the arrow discharged, the ratchets are lifted clear of the cogs on the stock of the engine.   This allows the trough to be slid forward to its first position as shown in A, B, fig. 8, p. 21.   It is then ready to be drawn back again for the next shot.

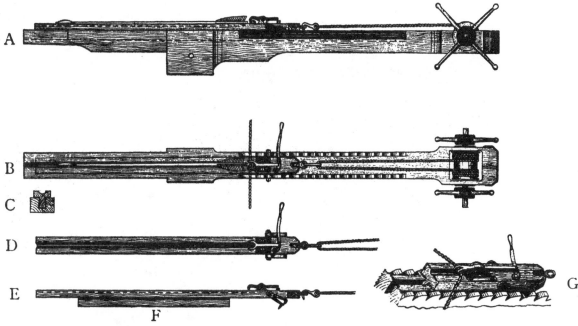

FIG. 8.—THE MECHANISM OF THE STOCK OF AN ARROW-THROWING BALISTA.

D.   Surface view of the trough, with the trigger and catch for the bow-string.

E.   Side view, showing the keel (F) which slides along the slot cut in the surface of the stock as the trough is drawn back by the windlass.

G.   Enlarged view of the solid end of the trough.   This sketch shows the catch for the bow-string, the trigger which sets it free, the ratchets which engage the cogs on the sides of the stock, and the slot cut in the stock for the dovetailed keel of the trough to travel in.

⁂

Balistas were constructed of different sizes for the various purposes of siege and field warfare.   The smallest of these engines was not larger than a heavy crossbow, though it more than equalled the latter in power and range.

The small balistas were chiefly used for shooting through loopholes and from battlemented walls at an enemy assaulting with scaling ladders and movable towers.

The largest had arms of 3 ft. to 4 ft. in length, and skeins of twisted sinew of 6 in. to 8 in. in diameter.

Judging from models I have made and carefully experimented with, it is certain that the more powerful balistas of the ancients could cast arrows, or rather feathered javelins, of from 5 to 6 lbs. weight, to a range of from 450 to 500 yards.

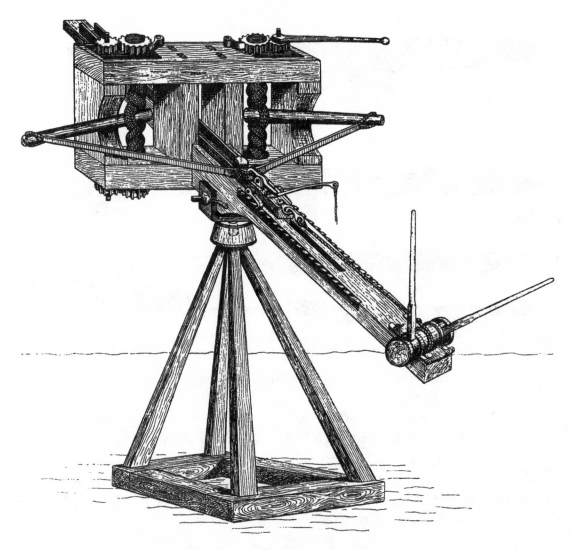

FIG. 9.—BALISTA FOR THROWING STONE BALLS.   Approximate scale : ½ in. = 1 ft.

This engine is here shown with its bow-string only slightly drawn along its stock by the windlass.

It will be seen that this engine is almost identical in construction with the one last described.   (Fig. 7, p. 19.)

The difference is that it propelled a stone ball instead of a large arrow.

The ball was driven along a square wooden trough, one-third of the diameter of the ball being enclosed by the sides of the trough so as to keep the missile in a true direction after the bow-string was released.

The bow-string was in the form of a broad band, with an enlargement at its centre against which the ball rested.

The description given of the mechanism and management of the engine for throwing arrows can be applied to the construction and manipulation of this form of balista, which was also made of large and small dimensions.

Small engines, with arms about 2 ft. in length and skeins of cord about 4 in. in diameter, such as those I have built for experiment, will send a stone ball, 1 lb. in weight, from 300 to 350 yards.

There is little doubt that the large stone-throwing balista of the Greeks and Romans was able to project a circular stone, of 6 to 8 lbs. weight, to a distance of from 450 to 500 yards.[1]

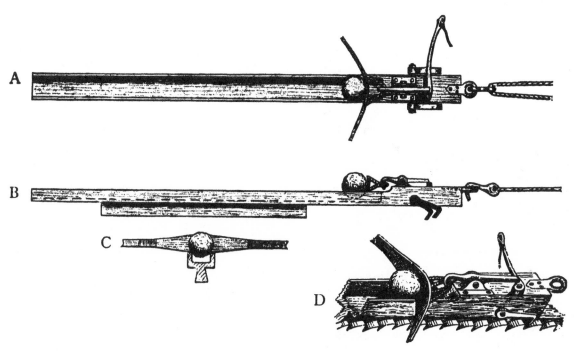

FIG. 10.—THE SLIDING TROUGH OF THE STONE-THROWING BALISTA.

A. Surface view, with the stone in position

B. Side view, with the stone in position.

C. Front view of the stone as it rests in the trough against the enlarged centre of the bow-string.

D. Enlarged view of the solid end of the sliding trough. This sketch shows the ball in position against the bow-string; the catch holding the loop of the bow-string, and the pivoted trigger which, when pulled, releases the catch. One of the pair of ratchets which engage the cogs on the sides of the stock,

---

[1] The balls used by the ancients in their catapults and balistas were often formed of heavy pebbles inclosed in baked clay, the reason being that balls made in this way shattered on falling and hence could not be shot back by the engines of the enemy. The balistas for throwing arrows, and those employed for casting stones, were fitted with axles and wheels when constructed for use in field warfare. (Pages 260, 273, 300, *The Crossbow*.)

as the trough is drawn back by the windlass to make ready the engine, is also shown. The trough has a keel to it, and slides to or fro along the stock in the same manner as in the arrow-throwing balista. (Fig. 7, p. 19.)

Compare with figs. 7, 8, pp. 19, 21, for further explanation of details.

———————————

For a detailed account of the history and effects in warfare of catapults, balistas, and other ancient projectile engines, refer to Chapters LII, LIII, LIV, and LVIII of 'The Crossbow.'

A TREATISE
ON TURKISH AND OTHER ORIENTAL BOWS
OF MEDIÆVAL AND LATER TIMES

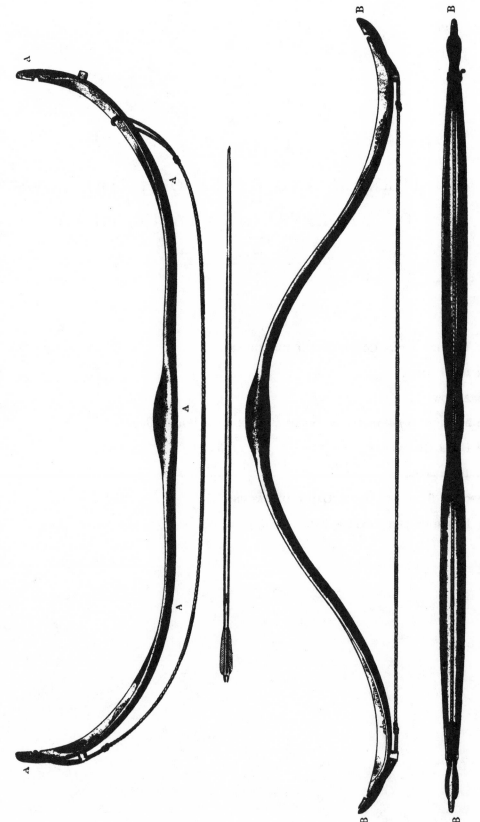

FIG. I.—TURKISH REFLEX COMPOSITE BOW UNSTRUNG AND STRUNG, AND ITS FLIGHT ARROW.

LENGTH of bow, measured, before it is strung, from end to end along its outer curve with a tape, 3 ft. 9 in. (AAAAA fig. 1, opposite page.)

Span of bow, measured between its ends when strung, 3 ft. 2 in. (BB fig. 1.)

Length of bow-string, 2 ft. 11 in.

Greatest width of each arm of bow, $1\frac{1}{8}$ in.

Thickness of each arm, at a distance of 6 in. from the centre of the handle of the bow, $\frac{1}{2}$ in.[1]

Circumference of each arm, at a distance of 6 in. from the centre of the handle of the bow, 3 in.

(The arms of the Persian, Indian, and Chinese composite bows have a width of from $1\frac{1}{2}$ to 2 in.; and though the span of these bows, when strung, is from 4 to 5 ft. and more, they do not shoot a light arrow nearly so far as the shorter, narrower, and in proportion far stronger and more elastic Turkish ones.)

The strength of the bow, or the weight that would be required on the centre of the bow-string to pull it down from the bow to the full length of the arrow, is 118 lbs. (This is without taking into account the additional two or three inches the point of the arrow should be drawn within the bow along the horn groove.)

Weight of bow, avoirdupois, $12\frac{1}{2}$ oz.

Though I have carefully examined over fifty of these small Turkish bows, I have never seen one that exceeded $1\frac{1}{4}$ in. in width at its widest part, or if measured with a tape along its outer curve, when unstrung (AAAAA, fig. 1), was over 3 ft. 10 in. in length. Bows that are 4 or 5 in. longer than the dimensions here given are invariably of Persian or Indian manufacture, and are very inferior in the elasticity that is requisite for long-distance shooting, though in decoration and construction they often closely resemble Turkish bows.

---

[1] In the very powerful bows, such as the one shown in Fig. 15, p. 21, the thickness at these parts is from $\frac{5}{8}$ to $\frac{3}{4}$ in.

The bow is chiefly constructed of very flexible horn and sinew.   These materials were softened by heat and water and then longitudinally glued to a slight lath of wood, varying from ⅛ to ¼ in. in thickness (except where it formed the handle of the bow), and from ½ to 1 in. in width.

This strip of wood formed the core or mould of the bow, and extended at each of its ends for 3 in. beyond the strips of horn and sinew that were fixed on its opposite sides, and which slightly overlapped it.   (Fig. 2, p. 5.)   The projecting ends of the wooden strip were enlarged so as to form the solid extremities of the bow in which the nocks for the bow-string were cut. (CC fig. 3, p. 6.)

The two curved horn strips, which in part comprised the arms of the bow (on its inside face when it was bent), were cut from the horn of a buffalo or an antelope, and average about ¼ in. in thickness.

The thicker ends of these pieces meet at the middle of the handle of the bow and their tapered ends extend to within 3 in. of its wooden points. (EE fig. 3, p. 6.)

The sinew that represents the back of the bow is from the great neck tendon of an ox or stag.   This was probably shredded longitudinally, and, after being soaked in elastic glue, compressed into a long flat strip about ¼ in. thick, which was first moulded in a pliable state to the wooden core and then glued to it.   It thus formed the back of the bow when it was bent. (DDD fig. 3, p. 6.)

The bark of the cherry-tree, or thin leather or skin, was next glued over the sinew to preserve it from injury and damp.   The horn parts, or inner face of the bow when it was strung, were not covered with bark or skin, a feature of the Turkish bow that, together with its small size, distinguishes it from the bows of India and other Oriental countries.[1]

In the best Turkish bows this outer coating of bark, leather, or skin, was lacquered a brilliant crimson and elaborately decorated with gold tracery, the date of the bow being always placed at one of its ends and the name of its maker at the other.

The horn and sinew (the materials which really form the bow and give it its power and elasticity) may be likened to a tube, the small centre of which is filled with wood.   (Sections, fig. 2, opposite page.)

---

[1] Though the horn strips which form the belly, or inner surface when it is strung, of a Chinese or a Tartar bow, are neither covered nor decorated, the great size of these weapons easily distinguishes them from those of Turkish manufacture.   (Fig. 13, p. 16.)

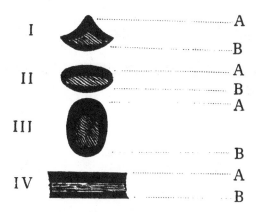

FIG. 2.—SECTIONS OF A TURKISH BOW.
Half full size.

I.   Section of bow at 6 in. from one of its ends.

II.   Section of bow at half-way between the centre of its handle and one of its ends.

III.   Section of bow at the centre of its handle, which is here thickly covered with sinew.

IV.   Longitudinal section of bow at half-way between the centre of its handle and one of its ends.

Light shading, AAAA.   The compressed sinew forming the back of the bow when it is strung.

Dark shading, BBBB.   The horn forming the inner surface of the bow when it is strung.

Lined centres.   The thin lath of wood to which the horn and sinew parts of the bow are moulded and fixed.

The thin wooden lath, in places only $\frac{1}{8}$ in. thick, bestowed no strength on the bow, as it was merely its heart or core to which the two curved strips of horn and the long band of sinew were glued.   (Fig. 3, p. 6.)

As it would have been very difficult and tedious to shape so fragile a lath in one length to suit the outline of the finished bow, this lath was always made in three pieces, which were fitted together at their joints and then secured with glue.   (Fig. 3.)

The middle piece formed the core of the handle of the bow and the other pieces the core of its limbs.   (Fig. 3.)

The extremities of the two outer pieces of the wooden core were enlarged to form the strong projecting points of the bow in which the nocks for the bow-string were cut.   (CC fig. 3.)

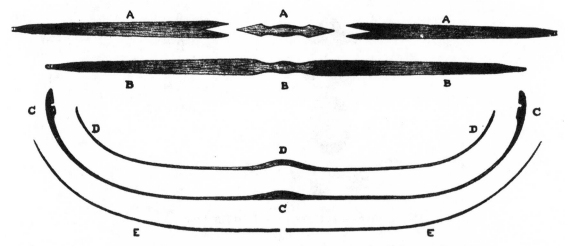

FIG. 3.—LONGITUDINAL PLANS OF THE PARTS OF A TURKISH BOW.

AAA.   The three pieces of thin wood that formed the core of the bow. Surface view.   (The two outer lengths of the core were steamed into a curve as shown in CCC.)

BBB.   The pieces glued together.   Surface view.

CCC.   The pieces glued together.   Side view.

DDD.   The strip of sinew that was glued to the core, and which formed the back or outer surface of the bow when it was reversed and strung.

EE.   The two strips of naturally curved horn that were glued to the core, and which formed the belly or inner surface of the bow when it was reversed and strung.

THE BOW-STRING

THE main part of the bow-string was composed of a skein of about sixty lengths of strong silk and was ingeniously knotted at each of its ends to a separate loop, formed of hard and closely twisted sinew.   A loop and its knot is shown in fig. 4, opposite page.

These loops could not fray or cut, as would occur if they were made of silk, and they fit into the nocks of the bow.   The loops rest, when the bow is strung, upon small ivory bridges (fig. 1, p. 2) which are hollowed out to receive them, and which, in this way, assist to retain the bow-string in its place.   Though these little bridges are not always present on Turkish bows, they are invariably to be found on those of Persian, Indian or Chinese construction, their greater length requiring the assistance of bridges to keep their bow-strings in a correct position.

FIG. 4.—ONE OF THE LOOPS OF HARD AND CLOSELY TWISTED SINEW WHICH ARE KNOTTED TO EACH END OF THE MIDDLE PART OR SKEIN OF A TURKISH BOW-STRING.

Scale : Half full size.

I. A loop and its knot as first formed on one end of the skein of the bow-string.

II. The loop drawn up, but not tightened.

III. The loop drawn up tight and its loose ends secured.

As shown in III, the projecting ends of the length of sinew which forms the loop are cut off to within a third of an inch of the knot. They are singed at their extremities, so as to form small burrs which prevent the short length of strong silk, which lashes them together, from slipping off.

The ends of this last small lashing are placed beneath the wrapping of silk to be seen on the skein near the knot in III.

In this way the knot of the loop is rigidly secured against any chance of drawing when the bow is in use.

(The bow-strings of all Oriental bows, with the exception of the Tartar and Chinese, were made as above described.)

THE ARROW

LENGTH of arrow, 25½ in. to 25¾ in.

Weight of arrow, avoirdupois, 7 drs., or equal to the weight of two shillings and a sixpence.

The balance of the arrow is at 12 in. from the end of its nock.

Shape of arrow, 'barrelled,' and much tapered from its balancing-point to its ends : its sharp ivory point being only ⅛ in. in diameter (where it is fitted to the shaft) and ¼ in. in length.

The part of the shaft to which the feathers are attached is $\frac{3}{16}$ in. in diameter, and the centre of the shaft $\frac{5}{16}$ in.

Though I have carefully measured and weighed about two hundred eighteenth-century Turkish flight arrows, I have scarce found a half-dozen that were $\frac{1}{8}$ in. more or less than from $25\frac{1}{2}$ in. to $25\frac{3}{4}$ in. in length, or that varied by even as little as $\frac{1}{2}$ dr. from 7 dr. in weight. In regard to their balancing-point these arrows are equally exact, as this part is invariably from $11\frac{1}{2}$ in. to $12\frac{1}{2}$ in. from the nock.

It is evident that the old Turkish flight arrow was made to a standard pattern that experience showed was the best for long-distance shooting.

The light and elegantly shaped wooden nock of an old Turkish arrow (fig. 5) is quite unlike the clumsy horn nock of the modern European one.

The latter cannot withstand the recoil of the Turkish bow and soon splits apart, though in the thousands of times I have discharged Turkish arrows I have never known one to split at the nock.

It will be noticed that the shape of the Turkish nock—with its narrow entrance that springs apart to admit the bow-string and then closes again— enabled an archer, even on horse-back, to carry an arrow ready for use on the string of his bow.

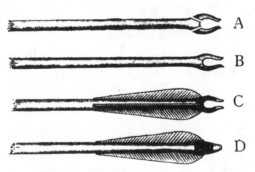

FIG. 5.—THE CONSTRUCTION OF THE NOCK OF A TURKISH ARROW.

Scale : Half full size.

A. The butt end of the arrow, with the projecting wooden halves of the nock shaped and ready to be glued to the shaft.

B. The halves of the nock glued to the shaft.

C, D. The feathers glued to the shaft.

The feathers (3) of a Turkish flight arrow, though stiff, are as thin as paper, and are $2\frac{1}{2}$ in. long and $\frac{1}{4}$ in. high near the nock. They were often made of parchment.[1]

The dark band of shading to be seen round the nock in C and D is a wrapping of fine thread-like sinew. This sinew, after being soaked in hot glue, was wound to a thickness of about $\frac{1}{32}$ in. all over the nock and it thus held the halves of the latter securely to the shaft.

When dry, the wrapping of sinew was cut out where it crossed the opening for the bow-string. It nevertheless gave a great increase of strength to the thin projecting halves of the nock, as it covered them on their outer surfaces

[1] Parchment feathering increases the range of a flight arrow by at least thirty yards. The reason of this is, that parchment is so thin and smooth that it offers very slight frictional resistance to the air, whilst at the same time it is much harder, as well as much more unyielding, than feather.

with a sheathing that was very tough and elastic, and as smooth as glass to the touch. This wrapping was, of course, applied before the feathers were glued on.

So careful were the Turks in the construction of these arrows, that even the halves of their nocks were made from wood with a natural curve to suit the finished outline. It is possible, of course, they would not otherwise have withstood the violent shock of the released bow-string. It may be said that every inch in length of a Turkish bow or arrow was named in a manner that could be recognised or referred to. In a general way the parts of an arrow were known as follows :—

The enlarged centre	The stomach.
From the centre to the point. . .	The trowser.
From the centre to the nock . . .	The neck.

THE METHOD OF STRINGING A TURKISH, PERSIAN OR INDIAN BOW

IN these days no person I have ever heard of can string a strong Turkish bow—diminutive as this weapon is—without much personal assistance, or else by mechanical means, yet formerly the Turkish archer unaided could do so with ease.

This he achieved by a combination of leg and manual power. (Figs. 6 and 7, p. 10.)

With the longer reflex bows, the Chinese for instance, this operation is comparatively easy, as the hand can reach one end of the bow and draw it inwards for the loop of the bow-string to be slipped into the nock.

The Turkish bow, being so short, necessitates a great effort of strength on the part of the archer to bend it between his legs and, at the same time, stoop down to fit the bow-string. From constant practice, the Turk of former days knew exactly how and when to apply the muscular force of leg and arm necessary to string his bow—a performance that no modern archer could accomplish with a bow of any strength.

Leg and manual force combined is the only possible method of stringing a strong reflex bow, unless mechanical power is utilised : it was the hereditary custom of the Orientals. In the operation, there is always the risk of twisting the limbs of the bow, from a lack of the great strength of wrist required to hold them straight during the stringing. If the limbs of the bow are given the slightest lateral twist as they are being bent, the horn parts are certain to splinter, and the bow is then useless and damaged beyond repair.[1]

The difficulty of reversing and stringing a very stiff bow with such a reflex curve that its ends nearly meet before it is bent may be imagined.

De Busbecq tells us that some of the Turkish bows were so strong that if a coin was placed under the bow-string at one end of the bow, as it was being strung, no one but a trained archer could bend the bow sufficiently to set free the coin so that it fell to the ground.

FIG. 6.   FIG. 7

Fig. 6 shows an Oriental reflex bow being gradually reversed preparatory to fitting on its bow-string.

Fig. 7 shows a similar bow when reversed sufficiently to fit its bow-string.

Though this illustration is from an ancient Greek vase, it will be noticed that in it the power of the leg and arm is applied in precisely the same way as in the more modern example given.

[1] The only safe method for a modern archer to adopt in order to string a powerful reflex bow is to use strong upright pegs, the size of tent pegs, inserted in smooth ground or in holes in a board, the bow resting during the process flat along the ground or board. Insert one peg against the inner face of the handle of the bow and then pull the ends of the bow back by degrees, placing a peg behind each of its ends as you do so to retain them in their acquired positions. The outer pegs can be shifted towards you as the bow is gradually bent, first at its one end and then at its other one. Finally, when the bow is fully bent the bow-string can be fitted across it from nock to nock and the pegs removed. To unstring the bow, grasp its extremities and, with the palms of the hands uppermost, bend it slightly across the knee, at the same time shifting with the thumb one of the loops of the bow-string out of its nock.

THE thin horn groove which the Turk wore on the thumb of his left hand when flight-shooting is shown in fig. 8.

This ingenious contrivance enabled the archer to draw the point of his arrow from 2 to 3 in. within the inner surface of his bent bow. He was thus able to shoot a short and light arrow, that would fly much farther than the considerably longer and heavier one he would have had to use if he had shot in the ordinary manner without the grooved horn.

The groove in the horn guides the arrow in safety past the side of the bow, when the bow-string is released by the archer.

The Turk, in fact, shot a short and light arrow from a very powerful bow, which he bent to the same extent as if he used an arrow 3 in. longer, with its proportionately increased size, weight, and frictional surface to retard its flight.

In the former case it will easily be understood that a much longer range could be achieved than in the latter.

Of this increase in length of flight conferred by the use of the grooved horn, the following experiment is conclusive evidence.

I lately shot from a Turkish bow twelve arrows, each arrow being three-quarters of an ounce in weight and $28\frac{1}{2}$ in. in length.

These twelve arrows were individually drawn to the head and the distance they reached averaged 275 yards.

I then reduced the same arrows to a length of $25\frac{1}{2}$ in. each, and to a weight of half an ounce each.

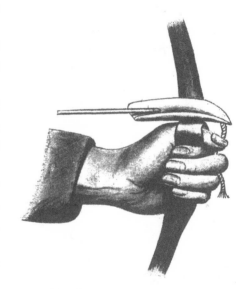

FIG. 8.—THE HORN GROOVE.

The bow is shown fully bent and ready for release, the point of the arrow being drawn back for a couple of inches inside the bow.

They were now shot from the same bow, over the same range and under the same conditions of weather, but their points were drawn $2\frac{1}{2}$ in. within the bow along a grooved horn. The distance they then travelled averaged 360 yards.

The Turk, as was the custom of Orientals, shot his arrow from the right-hand side of his bow, as shown in fig. 8, p. 11.[1]

The bow is here represented as fully bent, the point of the arrow being drawn back along the groove of the horn for a couple of inches within the bow.

The horn is attached to the thumb by a small leathern collar.

A short plaited cord of soft silk is suspended from the fore-end of the horn and is gripped between the fingers of the archer as he holds the bow.

This cord enables the archer to keep the horn in a level position on his hand.   It is fixed to a small strip of leather which is glued beneath the horn.

The horn is usually of tortoiseshell, very highly polished.   It is from 5 to 6 in. long, 1 in. wide, $\frac{1}{4}$ in. deep inside and $\frac{1}{16}$ in. thick.

It is slightly sloped from its centre of length to each of its ends, so that when the arrow is projected it touches the hard and smooth surface of the horn very lightly, and with, therefore, the least possible friction to retard its flight.

As the horn groove is only one-sixteenth of an inch thick, the arrow, as it is drawn back or shot forward, may be said to fit close against the side of the bow.

THE THUMB-RING.

THE Turk pulled his bow-string with a ring of ivory, or of other hard material, fitted on his right thumb.   (Fig 9, p. 13.)   Its manipulation is shown on p. 14.

It might be supposed that the strain of the bow-string on the ivory ring would cause the edges of the latter to injure the flesh and sinews of the thumb; this is not, however, the case in the least.

I find I can bend a strong bow much easier, and draw it a great deal farther, with the Turkish thumb-ring than I can with the ordinary European finger-grip.

The release to the bow-string which is bestowed by the small and smooth point [in Turkish "lip"] of the thumb-ring, is as quick and clean as the snap of a gunlock when a trigger is pulled, and very different in feeling and effect from the comparatively slow and dragging action that occurs when the release takes place in the European way from the leather-covered tips of three fingers.

---

[1] To discharge the arrow from the left-hand side of the bow, as is the custom in all European archery, the leather ring and the grooved horn will have to be fitted to the first joint of the forefinger.

The range of a flight arrow when shot from a bow by means of a thumb-ring is always much beyond that of an arrow shot with the three fingers in the usual manner.

With the thumb-ring the feathers of an arrow can be placed close to its nock, as the usual space of about 1½ in. need not be left on the shaft at the butt-end lest the fingers holding the bow-string should crush the feathers of the arrow—a precaution that is necessary in all European archery.

There is no doubt that the closer to the nock the feathers of an arrow can be fixed, the farther and steadier it will travel.

The handle of an English bow, or of any other bow that is loosed with the fingers, is placed below its centre so that the arrow can be fitted to the middle of the bow-string, a point which is just above the hand of the archer as he grasps the bow.

A bow held below its centre can never be pulled really true, the limb below the handle being shorter than the one above it.

In a Turkish bow the handle is in its exact centre of length, and the projecting point, or lip, of the thumb-ring engages the bow-string close to its centre.

For these reasons the bow is equally strained, each of its limbs doing its proper share of work in driving the arrow, an advantage that is very noticeable in flight-shooting, and would probably also be at the target. In the method of loosing used in modern times the bow-string lies across the three middle fingers, its outline, where the arrow is nocked on the string, taking the form of two angles connected by a straight line 2½ to 3 in. in length.

With the thumb-ring the bow-string is drawn back to one sharp angle close to the apex of which the nock of the arrow is fitted, so that every part of the string is utilised in driving the arrow. (Fig. 12, p. 14.)

FIG. 9.—THE TURK-ISH THUMB-RING. (Scale, half full size.)

The ease with which a strong bow can be drawn with the thumb-ring, and the entire absence of any unpleasant strain on the thumb, is remarkable. This proves how effective the Oriental style of loosing a bow-string was, compared with the one practised by European archers.

The ring was usually of ivory, its edges being round and smooth where they came in contact with the skin of the thumb.

A covering of soft leather was sometimes glued all over the sloping outer surface of the projecting lip of the ring.

The leather assisted the archer to hold the ring firmly with his forefinger, so that it could not slip under the strain of pulling back the bow-string. The projecting lip of the ring bestowed the leverage which enabled the archer to draw the bow-string of a powerful bow.

Thumb-rings of silver or of agate were often permanently worn by Turkish archers of position, both for ornament and for use.

These rings were finely polished and frequently inlaid with gold.

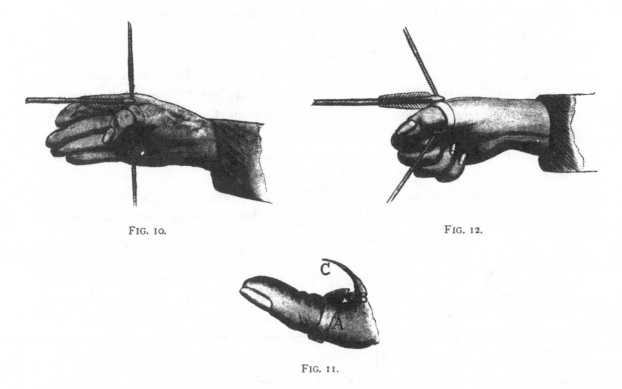

FIG. 10.   FIG. 12.

FIG. 11.

THE TURKISH THUMB-RING AND ITS MANIPULATION.

Fig. 10. The position of the hand when the arrow is first fitted to the bow-string, the latter being hitched behind the lip of the thumb-ring. The nock of the arrow should be close against the lip of the ring, and hence within about an eighth of an inch of the angle formed in the bow-string when it is fully drawn, as shown in fig. 12.

Fig. 11. View of the thumb, with the ring, A, in position preparatory to closing the forefinger and thumb.

[B. Section of the bowstring as hitched behind the projecting lip of the ring.

C. The base of the forefinger, or the part of it which presses tightly over the sloping surface of the lip of the ring, in front of the bow-string, when the bow is being bent.]

Fig. 12. The base of the forefinger pressed against the ring, the hand closed, and the bow-string and arrow being drawn back by the thumb-ring.

It should be noted that no part of the hand is utilised in holding the ring and in drawing the bow-string, except the thumb and the base of the forefinger.

When the pressure of the forefinger is taken off the ring (by separating this finger and the thumb) the bow-string instantly pulls the lip of the ring slightly forward, and at the same moment slips off it with a sharp 'click.'

The archers of other Oriental nations besides the Turks employed thumb-rings of various shapes and dimensions to suit the construction of their bows, bow-strings and arrows.   All thumb-rings were, however, more or less similar, and were all used in the manner I have described.

It is, indeed, impossible to shoot an arrow by means of a thumb-ring except as I have shown, and as a very short practical trial will prove.

If the ring is applied in any other way it either flies off the hand when the bow-string is released; the thumb is injured; or the bow-string escapes from its hold when only partially drawn.

In one of the Turkish manuals on Archery translated by Baron Purgstall (p. 22), many illustrations are given of the construction of the Turkish composite bow, but, unfortunately, minor details are omitted, though doubtless they were common knowledge when the Ottoman author wrote.

Without these details the correct formation of the bow cannot be ascertained. The chief omissions are (1) The composition of the very strong and elastic glue with which the parts of the bow were so securely joined, (2) The treatment of the flexible sinew which formed the back of the bow—whether, for instance, it was glued on in short shredded lengths or was attached in one solid strip.

All we know is that the sinew was taken from the *Ligamentum Colli* of an ox or stag, a very powerful and elastic tendon which contracts or expands as the animal raises or lowers its head to feed or drink.

When the sinew which comprises the back, or outside when it is strung, of a Turkish bow—however old it be—is dissolved in hot water, it disintegrates into hundreds of short pieces of from 2 to 3 in. long and about $\frac{1}{8}$ in. in diameter, each as ductile as indiarubber and almost unbreakable by hand.

The component parts of a Turkish bow, consisting of a thin strip of horn, one of wood and another of sinew (fig. 3, p. 6), are so pliable when separated that they can almost be coiled round the fingers, though if the same pieces are glued together they form a bow of unrivalled strength and elasticity.

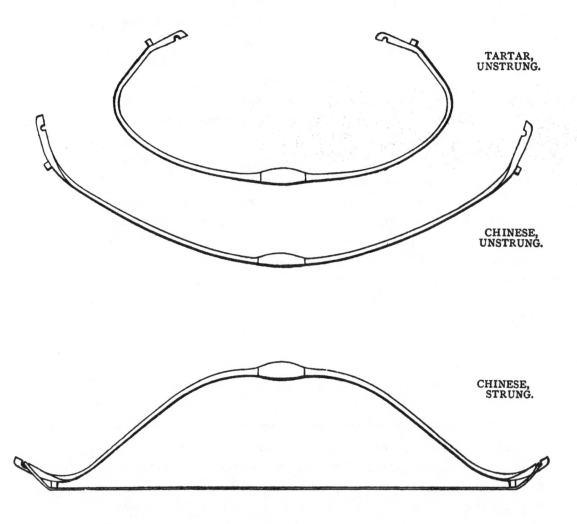

TARTAR,
UNSTRUNG.

CHINESE,
UNSTRUNG.

CHINESE,
STRUNG.

Scale : One inch = one foot.

FIG. 13.

FIGS. 13, 14. THE COMPARATIVE DIMENSIONS OF REFLEX COMPOSITE
BOWS OF VARIOUS NATIONS.—The structure of all these bows is similar in that
they are composed of sinew, wood and horn, *i.e.* sinew on the back of the bow,

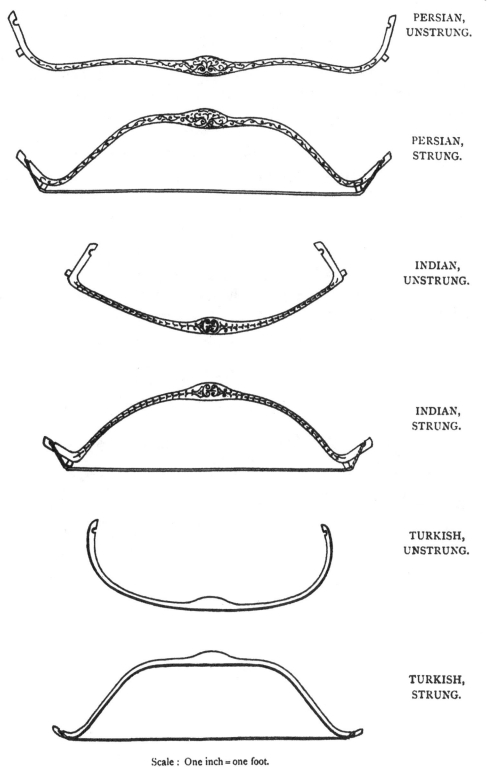

PERSIAN,
UNSTRUNG.

PERSIAN,
STRUNG.

INDIAN,
UNSTRUNG.

INDIAN,
STRUNG.

TURKISH,
UNSTRUNG.

TURKISH,
STRUNG.

Scale : One inch = one foot.

FIG. 14.

naturally curved horn on its inner face, and a thin core of wood between the horn and sinew.

Though the range of the Turkish bow—whether with a flighting or with a war arrow—far exceeds that of the other bows depicted, yet the Persian and Indian weapons are capable of shooting to a long distance, certainly much farther than any European longbow.

The great Chinese or Tartar bow requires a very long arrow, which from its length is, of necessity, a heavy one with a thick shaft. It cannot be propelled, as a result, farther than from 250 to 260 yards. One distinctive feature of Chinese, Tartar, Persian or Indian bows is the formation of their bow-strings. These are invariably from $\frac{1}{4}$ to $\frac{5}{16}$ in. in thickness, and are always closely wrapped round, from end to end, with soft cord or coloured silk of about the substance of worsted.

The Turkish bow-string is $\frac{1}{8}$ in. thick, and is merely served round with fine silk for 3 in. at its centre of length, with three or four shorter lashings at intermediate points.

THE LENGTHS OF THE ARROWS FORMERLY USED IN WARFARE WITH THE BOWS
GIVEN IN FIGS. 13 AND 14.

Chinese or Tartar bow	3 ft.
Persian	2 ft. 8 in.
Indian	2 ft. 6 in.
Turkish [1]	2 ft. 4$\frac{1}{2}$ in.

[1] The long Turkish war arrow was drawn to the head as in an ordinary bow. The grooved horn was only used with the short and light flight-arrow.

THE AUTHOR SHOOTING WITH
A TURKISH BOW.[1]

IN 1795 Mahmoud Effendi, Secretary to the Turkish Ambassador in London, shot a $25\frac{1}{2}$ - in. flight arrow 480 yards. The bow he used is similar to the one shown in fig. 1, p. 2, and is now preserved in the Hall of the Royal Toxophilite Society, Regent s Park.

Mahmoud Effendi accomplished this feat—which was carefully verified at the time—in the presence of a number of well-known members of the Toxophilite Society of the day, including Mr. T. Waring, the author of a work on Archery.

Joseph Strutt, the historian, was also a spectator, and describes the incident in his book entitled 'The Sports and Pastimes of the People of England.'

It is beyond question that in the seventeenth and eighteenth centuries, with bows precisely similar to the one shown in Fig. 1, but of much greater power, flight arrows were shot from 600 to 800 yards by certain famous Turkish archers.

The achievements of these celebrated bowmen were engraved on marble

[1] There are many country residences in England at which the author has made very long shots with a bow and arrow, and where trees have been planted to mark the distances. Among others ; Glynllivon Park, Carnarvon ; Broomhead Hall, Sheffield ; Onslow Hall, Shrewsbury ; Norton Priory, Runcorn ; The Hendre, Monmouth, and Harpton Court, New Radnor, may be named.

columns erected at the ancient archery ground near Constantinople, and these records are still in existence.[1]

The only trustworthy evidence of unusual ranges attained with the English longbow is as follows :

1798.	Mr. Troward	.	.	.	.	340 yards.
1856.	Mr. Horace Ford	.	.	.	308	,,
1881.	Mr. C. J. Longman	.	.	.	286	,,
1891.	Mr. L. W. Maxon	.	.	.	290	,,
1897.	Major Joseph Straker .	.	.		310	,,

It is not probable that the English bowmen of mediæval days were able to shoot the arrows they used in warfare farther than from 230 to 250 yards. Nor is it likely that they could send flight arrows to longer ranges than those given above, as heavy yew bows, strong as they may have been, were unsuitable for the purpose.[2] It was from their great elasticity, as much as from their strength, that composite bows derived their wonderful power.

When, too, the composite bow was strung, its bow-string was much more taut than was that of any European bow, as the latter was merely bent out of a straight line, whilst the former was bent from a sharp reflex curve, which it was always striving to resume when in use.

Though many nations formerly used composite bows of horn and sinew, no people attained such dexterity in their manipulation, or constructed them of such marvellous power and efficiency, and at the same time so small, elegant and light, as did the Turks.

It should not be supposed, however, that because these bows were so diminutive in size, they were mere playthings for shooting a flight arrow to an immense range. They were powerful weapons of warfare, and, as I have proved in practice, those of only moderate power are capable of sending an iron-shod arrow weighing 5s., or one ounce, to a distance of 280 yards. Bows that could shoot a flight arrow 600 yards, and more, would certainly be able to drive an ounce arrow 360 to 400 yards—or much farther than was possible with the old English longbow and its war shaft.

I have obtained with much difficulty during the last few years about a score of composite bows of Turkish manufacture from various parts of the Ottoman Empire. Not more than three or four of these have, however, proved serviceable, owing to their age, as no bows of the kind have been made for over a hundred years, the art of their construction being long since neglected and lost.

[1] See *The Crossbow*, pp. 28, 29.
[2] In King Henry IV., Second Part, Act III., Scene 2, Shakespeare makes Shallow exclaim of Double that the latter could shoot a flight arrow from 280 to 290 yards. In the time of Shakespeare (1564-1616) it was, therefore, considered a notable feat to send an arrow to this distance.

With the bow depicted in Fig. 1, I shot six arrows in succession to ranges exceeding 350 yards, the longest flights being 360, 365 and 367 yards. This public record was established July 7th, 1905, at an archery meeting held at Le Touquet, near Etaples in France. The ground selected for the trial was perfectly level; there was no wind, and the distances were accurately measured by several well-known members of the Royal Toxophilite Society who were present.

With the same bow I have, in private practice, thrice exceeded 415 yards, and on one occasion reached 421 yards.[1]

Though this bow is a powerful one for a modern archer to draw, it is a mere plaything compared with other Turkish bows of the same length, but of far greater strength, which I possess.

Some of the latter are so curved in their unstrung state that their ends nearly meet, and are so stiff, when strung, that I cannot draw them to more than

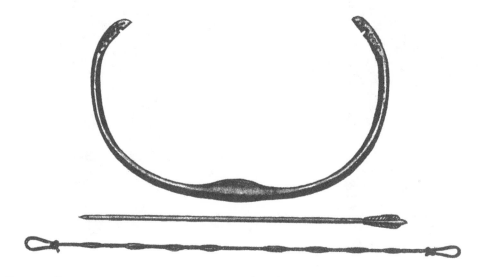

FIG. 15. SKETCH OF A VERY POWERFUL TURKISH BOW WITH ITS ARROW AND BOW-STRING.

half the length of a 25½-in. arrow. Fig. 15 shows a bow of this kind in my collection.

Such bows as these require a pull of 150 to 160 lbs. to bend them to their full extent, which quite accounts for the marvellous, but well authenticated, distances attained in flight-shooting by the muscular Turkish bowmen of bygone days.

Though 367 yards is a short range in comparison with that which the best Turkish archers were formerly capable of obtaining, it is, so far as known, much

---

[1] I presented this bow, and some of the arrows I used at Le Touquet, to the members of the Royal Toxophilite Society. These are now preserved in the club house of the Society in Regent's Park, the fine hall of which contains an unrivalled collection of archery implements and curiosities.

in excess of the distance any arrow has been shot from a bow since the oft-quoted feat of Mahmoud Effendi in 1795, p. 19.

Full corroboration of the wonderful flight-shooting of the Turks may be found in some treatises on Ottoman archery which have been translated into German by Baron Hammer-Purgstall (Vienna, 1851).

In his directions concerning the selection of suitable bows and arrows for the sport, one of the Turkish authors quoted by Purgstall writes : ' The thinnest

TURKISH CAVALRY SOLDIERS WITH THEIR BOWS.

From an illuminated Turkish MS. in the Sloane Collection, B.M., dated 1621, No. 5258.
These reproductions plainly show how small was the size of the bow formerly used in warfare
by Turkish soldiers.

and longest flying arrow has white swan feathers shaped like leaves,[1] and this arrow, with a good shot, carries from 1,000 to 1,200 paces.'

The orthodox length of a pace is thirty inches, and thus even 1,000 paces, or the lesser range mentioned, would exceed 800 English yards.

Augier Ghislen de Busbecq (1522-1592), a Belgian author and diplomatist, describes the Turkish archery he witnessed when ambassador to the court of Solyman, and the well-nigh incredible distances to which he saw arrows propelled.

---

[1] *Anglice*, Balloon feathers.

Full information to the same effect, with excellent diagrams, may be found in a Latin MS. on Turkish archery by J. Covel, D.D., Chaplain to the Embassy at Constantinople 1670–1676.[1]

Another treatise (in Turkish) entitled 'An Account of some famous Archery Matches at Bagdad (1638–1740), dedicated to the Governor of that city by the author, M. Rizai,[2] may also be consulted, as it gives the exact ranges of the longest-flying arrows.

It should be remembered that many years ago flight-shooting was a very popular recreation of the Turks, that every able-bodied man was a practised archer, and that every male child was trained to use a bow from the earliest possible age.

The origin of Turkish and other highly finished composite bows, and the approximate date when they were first used in sport and warfare, it is now impossible to determine. Bows that are undoubtedly of this kind and which are of excellent shape and design, are depicted on some of the most ancient pottery existent, and are also referred to in some of the oldest writings we possess.

For a full account of Ottoman archery and the extraordinary feats of Turkish bowmen, see pp. 27, 28, 29, 30, *The Crossbow.*

[1] MSS., B.M., 22911, folio 386.   [2] Sloane MSS., B.M., 26329, folio 59.